Guides to Professional English

T0198853

Series Editor
Adrian Wallwork
Pisa, Italy

For further volumes:
http://www.springer.com/series/13345

Adrian Wallwork

Presentations, Demos, and Training Sessions

 Springer

Adrian Wallwork
Pisa
Italy

ISBN 978-1-4939-0643-7   ISBN 978-1-4939-0644-4 (eBook)
DOI 10.1007/978-1-4939-0644-4
Springer New York Heidelberg Dordrecht London

Library of Congress Control Number: 2014939611

Printed on acid-free paper

Springer is part of Springer Science+Business Media (www.springer.com)

INTRODUCTION TO THE READER

Who is this book for?

This book is a guide to drafting and giving presentations and demos in a work or technical (but not research) environment.

It is intended for those who need to sell or explain their products and services, and / or to provide training.

The book is designed to help both those who have never done presentations before, as well as those whose English is already good (or who are native speakers) but who want to improve their presentation skills.

The focus is on language, rather than on the creation of slides from a technical/ artistic point of view (there are hundreds of sites on the web that can help you with this).

I hope that trainers in Business English will also find the book a source of useful ideas to pass on to students.

I work in research, is this book for me?

No. Although much of this book is relevant also for presenting research projects, a much better option for you is to read *English for Presentations at International Conferences* (Springer). Some chapters (with some modifications) are common to both books, specifically the chapters on preparation, pronunciation, bullets and Q&A (Chapters 4-9 and 15 of this book). However, *English for Presentations at International Conferences* contains specific chapters on how to structure the various parts of a presentation (Introduction, Methods, Results, Discussion, Conclusion) and on how to create a captivating beginning.

What will I learn from this book?

This book will help you to:

- prepare and practice a well organized, interesting presentation
- give effective demos and training sessions either on site or via audio/ video conference
- highlight the essential points you want the audience to remember
- avoid problems in English by using short easy-to-say sentences
- attract and retain audience attention
- decide what to say at each stage of the presentation
- improve your pronunciation
- learn useful phrases
- deal with questions from the audience
- overcome problems with nerves and embarrassment
- gain confidence and give a memorable presentation
- motivate your audience to listen and act on what you have said

How should I read this book?

This book is designed to be like a manual or user guide – you don't need to read it starting from page 1. Like a manual it has lots of short subsections, and is divided into short paragraphs with many bullet points. This is to help you find what you want quickly and also to assimilate the information as rapidly and as effectively as possible.

The first two chapters highlight the importance of doing presentations and how much you can learn by watching other presentations (e.g. on ted. com and YouTube). Chapters 3–10 cover all the preparation for your slides and for what you are going to say (including pronunciation and intonation). Chapters 11–15 cover how to conduct a presentation, demo and training session, either face to face or via video or audio conference. The chapters include how to answer questions, set tasks, manage the audience etc. Chapter 16 gives ideas on how to practice for your presentation, and Chapter 17 on how to improve your demo and training skills. The last three chapters contain lists of useful phrases.

All the chapters apart from Chapters 12–14, which are specifically on giving demos and training, are relevant to all types of presentations.

You can use the Table of Contents as a check list of things to remember.

Why are there no images of presentation slides?

This book is designed primarily to teach you what to say and how to say it.

The technologies for creating slides are constantly changing, so it makes more sense for you to see the most up-to-date slides by searching on the web. You can try the following sites:

prezi.com

google.com/drive

slideshare.net

presentationzen.com

authorstream.com

bbc.co.uk/worldservice/.../unit3presentations/1opening.shtml

glossary

audience, attendee, participant	I use these terms indifferently to mean the people you are talking to when you give a presentation or demo
demo	a presentation of a product or service in which the presenter demonstrates how the product or service works. Sometimes, I use demo and presentation indifferently
training session	a meeting in which one or more trainers teach participants how to do something. Such training often takes place through a demo
audio conference call	a telephone call between multiple people, without video
video conference call	a telephone call between multiple people, with video
video conference	a meeting with participants in multiple locations in which participants can see each other on a big screen

Other books in this series

There are currently five other books in this series.

CVs, Resumes, and LinkedIn
http://www.springer.com/978-1-4939-0646-8/

Email and Commercial Correspondence
http://www.springer.com/978-1-4939-0634-5/

User Guides, Manuals, and Technical Writing
http://www.springer.com/978-1-4939-0640-6/

Meetings, Negotiations, and Socializing
http://www.springer.com/978-1-4939-0631-4/

Telephone and Helpdesk Skills
http://www.springer.com/978-1-4939-0637-6/

All the above books are intended for people working in industry rather than academia. The only exception is *CVs, Resumes, Cover Letters and LinkedIn*, which is aimed at both people in industry and academia.

There is also a parallel series of books covering similar skills for those in academia:

English for Presentations at International Conferences
http://www.springer.com/978-1-4419-6590-5/

English for Writing Research Papers
http://www.springer.com/978-1-4419-7921-6/

English for Academic Correspondence and Socializing
http://www.springer.com/978-1-4419-9400-4/

English for Research: Usage, Style, and Grammar
http://www.springer.com/978-1-4614-1592-3/

INTRODUCTION FOR THE TEACHER / TRAINER

Teaching Business English

I had two main targets when writing this book:

- non-native speakers (business, sales technical)
- Business English teachers and trainers

My teaching career initially started in general English but I soon moved into Business English, which I found was much more focused and where I could quickly see real results. The strategies I teach are almost totally language-independent, and many of my 'students' follow my guidelines when writing and presenting in their own language. I am sure you will have found the same in your lessons too.

Typically, my lessons cover how to:

1. write emails

2. participate in meetings

3. make phone calls

4. socialize

and perhaps most difficult of all, how to do a presentation or demo. While I think I have mastered the first four, I always learn something new when watching a new client give a presentation or demo.

This book is thus a personal collection of ideas picked up over the last 25 years. It is not intended as a course book, there are plenty of these already. It is more like a reference manual.

I also teach academics how to present their work. In fact, some of the chapters in this book are based on chapters from *English for Presentations at International Conferences* (Springer).

How to teach presentations and demos

The reason I am not over keen on course books on Business English skills is that they assume you have 30–70 hours available to learn how to do something. In my experience, most of my 'students' learn that they have to do a demo a week or two before. They simply don't have time to follow a neatly structured course, but instead need a series of instant fixes.

My advice to you is to jump straight in and get your students to do their presentation or demo in front of you. Note down the areas where they are having the most problems and then just focus on fixing those.

I find that the biggest problems, apart from pronunciation/ intonation and difficulties in understanding questions from the audience, have nothing to do with English. Instead the problems are:

- no clear structure
- too much text in the slides
- reading the slides rather than interpreting them
- tendency to improvise and waffle
- no eye contact with audience
- no interaction with the audience
- no enthusiasm

In this book you can find various solutions to the problems above, and of course how to deal with English-related problems.

If you have more time available, then a good approach is to:

- find out what they like and dislike about other people's presentations (Chapter 1)
- help them to understand that just because they can see the faults in other people's presentations does not mean they will be able to recognize the same faults in their own presentations
- show them lots of presentations done by 'experts' e.g. on ted.com (Chapter 2)
- teach them how to write a script for their presentation and highlight the benefits of doing so (Chapter 4)
- focus on improving their pronunciation (Chapter 9)
- teach them how to teach (i.e. most of the skills you have as an ELT teacher, they will need if they have to give demos or training sessions)

I am lucky to have had a lot of experience in giving presentations myself (i.e. to promote my books at BESIG and IATEFL conferences), so that has helped me massively as I understand exactly how it feels to stand up in front of a group of strangers. If you can somehow get similar experiences, this will help you too.

Also, I find it helps a lot to take part in real demos. For example, I have recently been helping some IT developers to give training sessions over the telephone (with no video). So I took part in their audio conference calls as a passive participant. Had I not done so, I would have had no idea of the difficulties involved and nor would I have been able to come up with solutions.

Keep up to date with the latest technologies and techniques - see 9.18 for a suggestion to use a tablet or mobile phone during a presentation.

In summary, get involved with the company / companies where you teach. You will find it much more satisfying!

Contents

1 THE IMPORTANCE AND CHARACTERISTICS OF A GOOD PRESENTATION

1.1 Why do presentations and demos?

Presentations are opportunities to:

- establish yourself within the company as being an expert in your field

- show that you have good communication skills (i.e. to persuade, to inspire, to motivate) and thus have management potential - in most surveys of companies, presentation skills were rated as being as important (if not more so) than technical skills

- learn a lot about the topic you are presenting, it will help you understand your topic better, and it may help you to see the topic from a new perspective (i.e. that of the user rather than the developer / designer)

Any kind of training session or demo is like selling a product – you are trying to get your audience to do something, e.g. to use some product or service.

Doing the training well will avoid future problems for you (and the company) – in the sense you won't have to waste time answering customers' queries.

A. Wallwork, *Presentations, Demos, and Training Sessions,*
Guides to Professional English, DOI 10.1007/978-1-4939-0644-4_1,
© Springer Science+Business Media New York 2014

1.2   What kind of presentations do audiences like to see?

We tend to like presentations that:

- are professional and a delivered by someone who is credible and confident
- look like they were prepared specifically for us and make it immediately clear why we should be interested
- have clear slides, with minimal detail and helpful and / or entertaining images
- tell us interesting, curious and counterintuitive things
- don't make us work too hard to follow what is being said - two or three main points, lots of examples, and not too much theory
- are delivered in a friendly, enthusiastic and relatively informal way
- entertain us and interact with us

In addition, an audience who is watching a sales presentation or demo will want answers to the following questions:

- will their product / service be worth its high cost?
- how reliable are they and their product / service?
- what can they do for us that we can't already do?
- how will they improve our working life?

1.3 What kind of presentations do audiences NOT like to see?

We tend NOT to like presentations where the speaker:

- has clearly not practiced the presentation
- has no clear introduction, a confused structure and no conclusions
- appears to be talking to himself / herself rather than engaging with the audience
- reads the slides
- has a series of similar slides full of text and diagrams
- relies on animations
- fails to address the audience's interest and only sees things from his / her point of view
- is too technical, too detailed
- speaks too fast, speaks with a monotone, speaks for too long
- shows little interest in his / her topic
- does not provide opportunities for participants to ask questions
- has no scheduled breaks for longer presentations / demos

1.4 What constitutes a professional presentation?

A 'professional' presentation is one where you put the audience first. You think about how the audience would most like to receive the information you are giving.

The key to an effective presentation is that you have a few main points that you want the audience to remember, and that you highlight these points during the presentation in an interesting, and if possible, enthusiastic way.

The important thing is to be relaxed. To become more relaxed the key is to prepare well and concentrate on the content, not on your English. Your presentation is not an English examination - your English does not have to be perfect. Be realistic and don't aim for 100 % accuracy otherwise you will be more worried about your English than about communicating the value of your product / service.

1.5   How can I increase my confidence?

You may have had no previous experience in presenting or training. Your boss may have decided that you are the right person to give a demo or training session simply because you are the only person available (and not because you have the right knowledge or the right personality).

If you are not happy with being chosen to give a presentation, demo or training session, you have to work on yourself psychologically. Otherwise the demo may be a disaster both for you and your audience. Here are some strategies that might work for you:

- understand why giving demos and training sessions is important
- decide to believe (even if only temporarily) that what you are presenting is useful
- take on a different persona, i.e. the persona of someone who actually enjoys his / her work and enjoys explaining things
- relax and don't use your level of English as an excuse for doing a bad job
- appreciate that training people can be very rewarding and it might be something that you grow to enjoy

Although you may not be a born presenter, you will probably have one or more of the following qualities:

- an above average knowledge and considerable experience in your field
- a passion about what you do
- an ability to explain difficult technical things clearly
- an ability to find the exact answer to questions from the audience
- a professional look
- a sense of humor

Try to use these qualities to give yourself confidence and to show the audience that you know what you are talking about even if your English is not perfect.

A good presentation requires many skills that can only be learned over time. If in the past you did a bad presentation very probably it was because you had not prepared sufficiently. When you then have to do your second presentation you will have that bad memory of the first. It is important to put that bad experience behind you. Concentrate on getting it right the second time by preparing good content and then practicing it in front of as many people as you can.

1.6 What are the dangers of PowerPoint and other presentation software?

If you buy twenty tubes of paint you don't automatically have a painting. Likewise, if you create a set of PowerPoint slides (or slides using any form of template) you don't automatically have a presentation. You just have a set of slides.

A presentation is slides plus a lot of practice.

Try practicing your presentation without using any slides. If you find it difficult, it means you are relying too much on your slides.

PowerPoint templates encourage you to

1. create a series of similar looking slides

2. use bullets on every slide

3. have the same background

4. have a title for each slide

The first three can lead to a very tedious and repetitively visual presentation. There are a limited number of PowerPoint backgrounds, and most audiences will have already seen most of them. Try to invent your own background, or if not use a very simple background color (e.g. blue background with yellow text).

But the fourth, titles, is very useful. Titles are like a map for the audience guiding them through the presentation.

Having similar looking titles (i.e. same color, font, and font size) throughout the presentation should be enough to give it a sense of cohesion and consistency. This means that you can vary the other three - the look, the use or not of bullets, and have a changing background where appropriate.

1.7 How does product training compare to other types of training?

Think about past teachers you have had – why did you like or not like their style? Think about lessons you enjoyed at school or university, why did you enjoy them?

- level of interest
- new information
- examples given
- amount of direct involvement
- opportunities for practice
- opportunities to ask questions
- length of lesson

A presentation is really no different from a lesson. The audience is there to learn. They don't need to know everything that you know – only what is pertinent to them. They want to learn with the least amount of effort possible and in the shortest possible time. They want to enjoy the learning process.

2 LEARNING FROM OTHER PEOPLE'S PRESENTATIONS

2.1 TED.com

The best way to begin to learn how to do presentations is to watch and analyse presentations done by other people.

Subsections 2.2-2.6 analyse three presentations that you can download from TED - simply type in the name of the presenter and the title of their presentation.

I will highlight both the good and the bad aspects of these presentations. The idea is that then you can use the same techniques to analyse TED presentations as well as those of your colleagues, and of course your own.

On TED you can download the scripts of the presentations. This is not just useful for helping you to understand the presentation, but also to see how the speech is constructed.

If you are a Business English trainer (see Introduction to the Teacher), then you might find it useful to show extracts of these presentations in your lessons.

A. Wallwork, *Presentations, Demos, and Training Sessions,*
Guides to Professional English, DOI 10.1007/978-1-4939-0644-4_2,
© Springer Science+Business Media New York 2014

2.2 TED example 1) Jay Walker: English Mania

Jay Walker is head of Walker Digital and was named by *Time* magazine as one of the fifty most influential leaders in the digital age. In this 4-minute presentation Walker tells his native English speaking audience why their language has become so important and how it is being learned throughout the world.

LANGUAGE AND SPEED

Let's analyse the opening minute of his speech:

> Let's talk about manias. Let's start with Beatle mania: hysterical teenagers, crying, screaming, pandemonium. Sports mania: deafening crowds, all for one idea -- get the ball in the net. Okay, religious mania: there's rapture, there's weeping, there's visions. Manias can be good. Manias can be alarming. Or manias can be deadly.

> The world has a new mania. A mania for learning English. Listen as Chinese students practice their English by screaming it.

72 words. 10 sentences. 60 seconds. That's an average of 7.2 words per sentence - much less than 100 words per minute. Jay speaks incredibly slowly and clearly. Is he talking to a group of English learners? No, he is talking to people who speak English as well as he does and could probably still understand him if he spoke three times as fast. Yet Jay chooses to:

- use short sentences
- use simple language
- speak very slowly and clearly

Why? To ensure that his audience does not have to make any effort to understand him. Also, by using short sentences it helps him to:

- remember what he wants to say
- speak clearly without hesitation

Are all Jay's presentations delivered in such a clear way with a slow speed? No. Jay varies his speed according to the importance of what he is saying. In the introductory part of another of his presentations on TED ("Jay Walker's library of human imagination") he speaks far more quickly as in rapid succession he shows the audience a few amazing artifacts from recent history. But when he begins talking about the main topic - the printing press - his voice slows down and takes on a more animated quality. He really wants his audience to understand what he is going to say.

2.2 TED example 1) Jay Walker: English Mania (Cont.)

STRUCTURE

Does Jay launch straight into his topic? No. He introduces the theme i.e. manias, but not the key topic i.e. English. This gives the audience time to

- adjust from the previous speaker to this new speaker
- hear something interesting, relevant, but not crucial
- tune into Jay's voice

It also allows the presenter to settle his nerves.

TRICKS

If you watch Jay's presentations, you may notice two things. One, he doesn't smile much. Two, he has notes. Although he may not be the most charismatic presenter on the planet, he recognizes his own limitations. Even though he doesn't smile a lot, he is still interesting - he packs his presentations with weird and wonderful statistics (but always pertinent). OK, so he can't remember every word he wants to say but he is confident enough to know that it is perfectly acceptable, even at this level of venue, to take a quick look at his notes. Alternatively, in your hand can hold a tablet or mobile phone where you can upload your entire speech / slides (see 9.18).

You will also notice that his slides have no text. They are simply there to remind him what to say, and to help the audience follow what he is saying.

2.3 TED example 2) Aleph Molinari: Let's bridge the digital divide!

Five billion people don't have access to the Internet. Economist Aleph Molinari tells us what we should do about it. Like Jay Walker, Aleph Molinari is not the most dynamic presenter, he doesn't run around the stage entertaining us. But he does know how to inform us and how to bring important data to our attention.

EXAMPLES AND STATISTICS

Aleph immediately starts with concrete examples of victims of the digital divide. He then moves on to some statistics. He shows a slide with the number of people in the world: 6,930,055,154.

Why not 7bn? Because the length and exactness of the number emphasizes firstly the incredible number of people who live on our planet and at the same time that they are individuals. The long number also looks dramatic on the screen. But when he actually mentions the number verbally he says "nearly seven billion people" - there would be no point in reading the exact number. He then gives the number of people who are digitally included, which on the slide appears as 2,095,006,005. What he says is "Out of these, approximately two billion are digitally included, this is approximately 30 % of the entire world population, which means that remaining 70 % of the world, close to five billion people do not have access to a computer or the Internet .. five billion people, that's four times the population of India".

Aleph's technique is thus to:

1. show a statistic in a simple clear way (i.e. not along with several other distracting statistics)

2. talk about the statistic in three ways (first as a whole number, then as a percentage, then by comparison with India). Aleph thus offers his audience different ways of absorbing the information, his aim being to help them to really understand the true significance of the numbers involved

3. interpret the statistic by saying what the implications are

TEXT, BACKGROUNDS AND FONTS

Aleph's slides have a black background with a yellow font. They are incredibly clear. The majority of his slides that contain text only have one or two words. The slide with the most text, which is his first slide and contains a definition of the digital divide, contains 19 words. At least half his slides are just photographs, which support his speech. Essentially, the information contained on each of his slides can be absorbed in less than two seconds. This means that all the audience can listen to him with 100 %

2.3 TED example 2) Aleph Molinari: Let's bridge the digital divide! (Cont.)

of their concentration, rather than some of the audience reading the slides and some listening to him.

CRITICISMS OF ALEPH'S PRESENTING STYLE

I imagine that he is quite introverted. This reveals itself in the fact that he spends too much time (in my opinion) looking at the screen rather than the audience. Although he does try to emphasize his key words, his voice is rather monotonous. The combination of these two factors could lead to the audience losing interest.

However, Aleph compensates for this lack of dynamism by

- having a clear logical structure
- having excellent slides - clear, easy to follow
- being professional

This makes him in the audience's eyes very credible. Although they may not be entertained they will certainly be motivated to follow him and listen to what he has to say. And this also means that although his conclusion in itself lacks much impact (his voice does not sound very impassioned), as a whole his presentation will have a positive impact because he appears to the audience as being totally committed to his project and also very sincere.

2.4 TED example 3) Philippe Starck: Design and destiny

As you can read on his TED biography, Philippe Starck is a well-known French product designer. His designs range from interior designs to mass-produced consumer goods such as toothbrushes, chairs, and even houses. I have chosen to analyse him because he is a non-native English speaker with what most people might consider to be not a very good English accent. He is also the only presenter not to use any slides at all.

ENGLISH LANGUAGE: GRAMMAR, VOCABULARY, PRONUNCIATION

Philippe Starck is worth watching in order to prove to yourself that even if you don't have a good English accent it doesn't necessarily matter. His technique for dealing with his poor English is to immediately draw attention to it in a self-deprecating way by saying: "You will understand nothing with my type of English."

His pronunciation is terrible. At least 20 % of his first 100 words contain pronunciation mistakes (e.g. 'ere instead of here, zat instead of that, the u in usually pronounced like the u in under rather than the u in universe) and he consistently puts the stress on the wrong part of a multi-syllable word (e.g. comfortable, impostor). He makes a series of grammar mistakes: forgetting the plural s, using the wrong part of the verb etc.

But because the audience are interested in what he is saying rather than how he is saying it, his poor English skills are not a problem. In fact if you read the comments on his presentation, not one reference is made to his poor English. Instead many viewers simply write: Superb! Fantastic! Really the most brilliant talk I've heard on TED.

However, note that Starck does speak slowly. If he had spoken very fast, this poor accent would probably have interfered with the audience's ability to understand him.

NO SLIDES, GOOD BODY LANGUAGE

Philippe manages to hold his audience's attention for 17 minutes without using a single slide. He is able to do this not just because he is a dynamic person who obviously loves an audience, but also because he has interesting things to say which he presents with a new perspective.

Another technique for retaining attention, is that he moves around the stage. This means that the audience have to follow him with their eyes, and this small bit of physical effort keeps them more alert. In addition, he uses his hands, and often his whole body, to give meaning to what he is saying.

2.5 What can we learn from these three TED presentations?

What these presentations all have in common is that it is clear that the presenters were well prepared. The audience feel that they are being led forward in a logical progression and that the presenter has spent a considerable amount of time practising his / her presentation. This gives each presenter credibility in the eyes of the audience, and also helps to make their presentation memorable.

It is probably these two factors - credibility and memorability - that you should aim at. And you can achieve this through:

* uncomplicated language

* loud, clear, slow voice

* simple slides

* a clear logical structure

Clearly, none of the three presentations have anything to do with business or giving demos. At least not in terms of content. But the factors that I have highlighted (body language, structure, simple slides etc) are nevertheless very relevant to any type of oral presentation.

2.6 The benefits of TED

You can choose the topic of the presentations you want to watch by using TED's search engine, and you can also choose whether to have English subtitles on or not. The subtitles report every single word, and are particularly useful for seeing (not just hearing) how many words a presenter uses in a sentence. This highlights that the shorter the sentence is, the easier it is for the presenter to say, and the easier it is for the audience to understand.

You can see or download a full transcript (called 'interactive transcript') of the presentation in English, plus translations in several other languages. This means that you can note down any useful phrases that the speaker uses that you think you could use too.

By reading the transcript and listening to the presentation at the same time, you can also improve your pronunciation and intonation by trying to imitate the presenter. For more on pronunciation and intonation, see Chapter 6.

2.7 Dragon's Den: learn how to pitch your ideas to venture capitalists

Another great source of presentations is Dragon's Den. This reality show originated in Japan and features entrepreneurs pitching their ideas to venture capitalists (known in the program as 'dragons'). There are versions of this program in over 20 countries worldwide, with several editions in English. For more details of the programs see: http://en.wikipedia.org/wiki/Dragons'_Den. The Wikipedia site explains under the section 'Show format' how the show works and the various 'rules'. This is essential reading in order to get the best out of the show. You can also learn useful tips about presenting, about the dragons themselves etc by consulting the websites of the shows (e.g. bbc.co.uk / programmes).

The English language versions vary considerably. The UK version, produced by the BBC, is the most serious and business-like, and for non-native speakers is probably the easiest to understand. But the Canadian version is also great fun, where the Dragons interact much more with each other and the contestants seem to have whackier ideas.

You can find many of the shows on YouTube, and each version has many series, so there is an infinite amount of material to watch and learn from. However, unlike TED there are no subtitles or transcripts for download.

By watching Dragon's Den you will learn some useful tips for making a sales pitch or presentation. Here are my top five tips in no particular order:

1) YOU ARE AS IMPORTANT AS YOUR IDEA

Basically venture capitalists invest as much in the person (i.e. the presenter) as in the idea. This is very important for you to know as it underlines the importance of how you as a person / personality come across during your demo and presentation. This means that you have to be:

- credible, honest and transparent
- approachable (i.e. potentially easy to work with)
- prepared to listen to what the dragons say, rather than talking over their questions - the dragons all hate presenters who fail to listen carefully (see Point 3 below)
- appropriately dressed (some of the UK dragons seem quite obsessed by the entrepreneur's dress code, one says he would never invest in someone who is 'scruffy')

While they do appreciate entrepreneurs who are enthusiastic or passionate about their product or service, the dragons don't like people who are arrogant or aggressive.

2.7 Dragon's Den: learn how to pitch your ideas to venture capitalists (Cont.)

2) KNOW YOUR FACTS AND PREDICT ALL THE POSSIBLE QUESTIONS THAT THE DRAGONS MIGHT ASK YOU

The dragons are venture capitalists, therefore one of their top priorities is money - how much money has your company made in the last quarter, last three years etc, how much do you estimate it will turn over in the next year etc. Contestants who are unable to answer these questions rarely get an investment.

This highlights the importance of knowing your audience. What are their priorities? What questions are they likely to ask me?

3) LEARN HOW TO LISTEN AND HOW TO ANSWER QUESTIONS

The times when the dragons become the most irritated is when the entrepreneurs interrupt the dragons while they are speaking or making suggestions. This means that when you are explaining something and someone interrupts you and starts talking, you should not continue talking yourself. Try to answer their question calmly and clearly, without showing any signs of irritation. This is particular important when dealing with clients, or with people higher up the hierarchy than you - they will not appreciate your determination to continue talking and may decide that you are impolite or even arrogant.

If however it is clear that the others in the audience wish you to continue, then very politely interrupt the questioner and say that you will be happy to deal with their question in a minute.

4) BE CLEAR AND CONCISE

Contestants only have three minutes to make their initial pitch (a pitch is a rapid presentation giving the essential facts). This means that every word has to add value. And this means that you cannot afford to improvise. You must know exactly what you want to say. To be able to do this, you really need to prepare a script (see Chapter 4).

5) PUT A REALISTIC VALUATION ON YOUR BUSINESS

The dragons never invest in a business that they consider overvalued.

The lesson to be learned here, is that in your presentation or demo, you have to motivate your audience to listen to you. If you say anything that is clearly unrealistic or not viable for your audience, then they will quickly stop listening.

2.8 Get ideas about what to say at the various stages of your presentation: Google IO

If, for example, you are unsure of the best way to introduce yourself or a good way to end your demo, then watch how others have solved these problems. Google IO conferences take place every year, and you can see a variety of presenters from top managers to developers. These people are at the top of their game, so if you copy / adapt their techniques you cannot go wrong.

However, bear in mind that most of these people are technicians, not trained presenters - so even they do not deliver perfect presentations. Hopefully, you should find this reassuring!

These presentations are available on YouTube and elsewhere. Obviously, you can learn similar ideas and techniques from conferences held by Microsoft, Apple and other large companies.

2.9 Learn from Steve Jobs

Steve Job's was considered one of the world's most captivating communicators. There are many articles and presentations on the web that analyse Jobs's techniques. Here are just a few.

http://www.slideshare.net/asad.taj/steve-jobs-presentation-skills

http://www.youtube.com/watch?v=RHX-xnP_G5s&feature=related

http://www.youtube.com/watch?v=7ABFW6rv15g&feature=related

2.10   Analyze other people's slides

There are several sites on the Web where you can share slides, for example:

slideshare.net

myplick.com

authorstream.com/slideshows/

These sites are useful for seeing how other people in similar fields to yours create their slides. Watching these presentations should help you to understand that packing a presentation full of detail is not usually a good approach.

When you have watched five or six presentations on TED (or whatever site), write down what you remember about the content and about the presenter and his / her style. You will be surprised how little you remember about the information that was given. Repeat the same memory exercise a week later and you probably won't even remember how many presentations you watched. Instead, you will remember the impression that the presenter made on you and their style of presenting for much longer.

What this means is that there is no point filling your presentation with descriptions of complex procedures or masses of data, because the audience will simply not remember. What they will remember from that experience is their frustration in not being able to absorb the information that you gave them. Make sure you always give your audience a positive experience.

2.11 Assess other people's presentations

You can learn a lot from the presentations you watch. Use the assessment sheet below to decide which presentation styles you liked and why. Then you can perhaps think of ways to incorporate these aspects into your own presentations.

Also, analyse the audience's reaction. Is the audience attentive? Are you yourself attentive? Notice when and why the presenter starts to lose your attention. If you stop watching, at what point did you stop watching and why?

	THE PRESENTER TENDS TO DO THIS	RATHER THAN THIS
CORE FOCUS	clarifies the main point of the presentation immediately - it is clear to audience why they should listen	the main point only emerges towards the end - audience not clear where the presentation is going
PACE / SPEED	varies the pace i.e. speaks slowly for key points, faster for more obvious information; pauses occasionally	maintains the same speed throughout; no pauses
BODY LANGUAGE	eyes on audience, moves hands, stands away from the screen, moves from one side of the screen to the other	eyes on screen, PC, ceiling, floor; static, blocks screen
STRUCTURE	each new point is organically connected to the previous point	there are no clear transitions or connections
FORMALITY	sounds natural, enthusiastic, sincere	sounds rather robotic and non spontaneous
STYLE	narrative: you want to hear what happened next lots of personal pronouns and active forms of verbs	technical, passive forms
LANGUAGE	dynamic, adjectives, very few linkers (*also, in addition, moreover, in particular*, etc)	very formal, no emotive adjectives, many linkers
RELATION WITH AUDIENCE	Involves / entertains the audience - thus maintaining their attention	seems to be talking to him / herself not to the audience
TEXT IN SLIDES	little or no text	a lot of text
GRAPHICS	simple graphics or complex graphics built up gradually	complex graphics
ABSTRACT VS CONCRETE	gives examples	focuses on abstract theory
STATISTICS	gives counterintuitive / interesting facts	makes little or no use of facts / statistics
AT THE END	you are left feeling inspired / positive	you are indifferent

3 YOUR AUDIENCE AND PREPARING THEM FOR YOUR PRESENTATION

3.1 Don't begin with the preparation of your slides

Before you start preparing your presentation / demo you should:

- make sure you understand why you have been chosen to do the presentation

- check to see if someone in the company has already done the same or similar presentation

- decide if a presentation is really the best way to give this information to your audience (alternatives: training session / workshop, document)

- think about how much time it will take you to prepare the presentation – then multiply this by at least five

Answering the question *Why I am doing this presentation?* generally gives answers such as: *To tell the client about our new product* or *To show the client how the new product works.* Try to find a precise and limited objective. More specific would be *To convince the client to switch to our product.* Even more specific would be: *To highlight the three factors in our product that would convince the client to switch.*

Then think about these questions.

- Is the presentation / demo for training? Is it internal, for an old client, for a new client?

- Does the audience need to understand every word? Or is your aim just to leave the audience with a general impression?

- Is the content important? If at the end the audience has not understood every single thing you have mentioned, how big a problem is it?

- Is it the fact that the audience is simply attending that is the most important thing? Will they receive details at a later date?

- Length – 20 mins, 30, 60 or more?

- Time of day? Experts say that, where possible, the best time of week to schedule presentations is between Tuesday and Thursday, in the mornings between 09.00 and 12.00.

A. Wallwork, *Presentations, Demos, and Training Sessions,*
Guides to Professional English, DOI 10.1007/978-1-4939-0644-4_3,
© Springer Science+Business Media New York 2014

- One presentation or a series?
- Support documentation (handout, on web)?
- Interactive or not?

3.2 Find out about the audience

Your presentation needs to appeal to as many people in the audience as possible. You need to know:

- who they are: professional background (technical, sales, business), nationality and level of English, age
- why they are coming e.g. because they need info or because they are obliged to attend (in which case you will need to work harder to get their attention)
- what they have in common with you
- what they need, and whether they all need the same thing
- how up to date they are on the topic you are presenting
- how resistant they might be to what you are proposing
- how much time they have
- what will get their attention
- what will make them act on what you are going to tell them
- what message you want them to take away

3.3 Get someone to email you the list of attendees and their roles

The simplest solution to finding out who will be attending, is to request a list of participants. This list could include a lot of information that would be useful to you. Let's imagine you work for an IT company in the US which has an office in Delhi, India. You are going to do a presentation of a software product for your colleagues in Delhi. You could ask for the following table to be filled in.

FIRST NAME	FAMILY NAME	MALE (M) FEMALE (F)	POSITION IN THE COMPANY	KNOWLEDGE OF XXX	SPECIFIC REASONS FOR ATTENDING
Nikki	Settigere	F	senior developer	good	to improve advanced skills in xxx
Praveeen	Huria	M	junior developer	good	-
Vibhor	Kamatchi	M	marketing manager	low	to gain tips on main features of the product

From the above table you would learn a lot of information:

- their name (to learn why this is important see 11.3)

- their sex – from their name it may be impossible to understand what sex they are. Knowing their sex may be useful if your demo is being done over the telephone as it may help you to understand better who is speaking

- their position, their level of knowledge of the topic you are going to talk about, and their reasons for attending.

The table above highlights that they are not a homogeneous group, so you will need to tailor your demo to the various needs of the people involved. You will also need to discuss this non-homogeneity with the group at the beginning of your presentation. For instance, maybe Vibhor, the marketing manager, is only at the presentation in a passive role, so you don't need to make concessions to him in terms of the level of technical detail.

But you need to have this info before you start the demo.

So ideally, having such a table is very useful. Realistically, you are unlikely to be able to get such information. Not because the info is not available, but simply because people will not find the time to give it to you. So as a compromise at least aim to have the names of the participants, then you will know how many people are attending.

3.4 Find out the numbers of people attending

It is very different giving a demo / presentation to small groups than to large groups. With a small group (e.g. 3-5 people) you:

- will have a more intimate relationship
- can tailor the presentation more to their requirements
- can deviate and improvise more – again to match the audience's needs
- will probably be a little less nervous

But there are advantages of big groups too:

- they can be more dynamic and you can have more fun
- you can follow your planned structure more closely because there is probably less opportunity to improvise

Knowing the number of people will also help if you have to prepare any photocopies.

3.5 Use company websites, Google, LinkedIn and Facebook to find out more about the attendees

From the attendees' company website or from LinkedIn you should be able to see what role an attendee has in their company (e.g. are they business or technical?). You can also see where their 'Skills & Expertise' lie. It might also help you to have an indication of how old they are and also what sex they are (you may not be able to tell this from their name alone).

Look at their photos. This will make them seem familiar when you see the person face to face, and if you have large groups of people, the people will seem less intimidating.

From their pages on Facebook, you can learn something about what they like doing in their free time – you might find you have something in common. But don't mention to attendees any information that you have found out about them, as they may feel you have been prying (i.e. investigating without their permission) into their private life.

3.6 Prepare handouts for demos and training sessions

A handout is a document that you give (or email) to participants. A handout will considerably increase the audience's ability to remember what you say in the presentation. It will also:

* avoid them having to take notes
* allow attendees to follow you even if they have difficulty understanding what you are saying
* help them follow difficult definitions or explanations
* enable you to show the full diagram, chart, figure or table, of which you are only able to show a part on your slide
* enable you to provide your contact details

Here are some general rules about how to create a handout:

* keep the handout short to increase the chances of participants reading it
* have small screenshots of the slides from your presentation – make sure each slide is numbered (and in the same order as in your presentation)
* put extra information (text and figures) next to or below the relevant screenshot
* prepare a one-page summary to put on the front or back page.

3.7   Email the handout in advance

When you have to give attendees a lot of information, email the handout in advance. Tell them:

- the purpose of the handout

- to read it before the presentation

Here is a possible email.

Dear X

As you know, I will be giving you a presentation on X on Friday 10 March. Attached are some notes on Y. I would be grateful if you could read them before the presentation itself. They are only two pages.

In any case, you cannot assume that attendees will have read the handout before they come to the presentation. Assuming no one has read the handout, at the beginning of your presentation you can say:

I have prepared some notes on fundamental definitions of x, y and z. While we are wait-ing for the others to arrive, could you just have a quick look through them and ask me for any clarifications.

When you do the presentation, check whether the audience has read the handout. In any case, do not merely repeat the info in the handout:

This part of the presentation is actually contained in the handout. So I am just going to go through it very quickly. I will be giving you two examples – x and y. In any case, I will tell you which parts of the handout you need to look at in more detail later.

If you didn't have time to prepare a handout, or to print it, you can always put the presentation slides directly onto your website for the audience to download.

3.8 Decide when is the best time to give the audience the handout

If you give the handout to attendees at the beginning of the presentation this allows you to:

- refer to the handout during the presentation

- have any detailed diagrams that the audience might not be able to see clearly on the screen

This solution is particular useful for participants whose listening comprehension in English is low level and who otherwise might not be able to follow the presentation. Refer to the handout throughout the presentation so that the audience knows where you are up to. Tell latecomers where you are in the handout so that they can quickly review what you have said so far.

However, the audience may be distracted and browse through the handout rather than follow you. So you may decide to give them the handout at the end. If you decide on this option, announce it at the beginning. Tell them there is no need to take notes.

Handouts are also useful for people who have to leave the presentation early, or who have to go out to make a phone call and then come back in later.

3.9 For demos, put yourself in the audience's shoes

Unfortunately, many demos and training sessions tend to be prepared from the perspective of the presenter rather than the audience. A really effective presentation has to answer the questions that the audience themselves would like to ask. Also, it should really convince them that what you are offering is going to change their working lives significantly for the better.

Which is more is important for them – that they can customize the toolbar of a software application that you are going to present them or that they will really be able to keep their customers happy? The toolbar can be customized in virtually any application but customers are essential to a company, so keeping them happy is of major importance. So, only present those features that are:

- really useful for the trainees, i.e. that will make a real difference to their working lives
- not already standard (i.e. don't talk about features that are common to most other applications)

You will gain their attention if you

- plan your demo showing the benefits for them, rather than demonstrating the ingeniousness of the product / application from the designer's point of view
- give plenty of examples which clearly relate to their needs
- ask them pertinent questions
- tell them the positive things other clients have said about the application

It doesn't matter if you don't cover everything, the most important thing is to give a

- good impression
- a clear picture of the functions that you present

If you have thought about all these factors while preparing your presentation, you could ask the audience directly what would improve their working life. You then write their answers on the whiteboard as bullet points. As you go through the presentation you can tick the bullets that you have covered. This is a really effective way of showing that your presentation really meets their needs.

3.10 Be aware of cultural differences

In his book *Outliers*, Malcolm Gladwell, a writer at *The New Yorker* magazine and named as one of *Time* magazine's 100 Most Influential People, talks about cultural differences in the way we communicate and receive information. In Chapter 8 he makes three very interesting points:

1. many Asian countries are 'receiver oriented', this means it is the listener's task to interpret what the speaker is saying

2. the Japanese have much higher levels of 'persistence' than Americans. This means that the Japanese can stick to a task for much longer than their American counterparts – they have higher levels of concentration

3. our memory span is correlated to the time it takes in our language to pronounce numbers. Because the words for numbers in Asian languages are quicker to pronounce and are more logical (*ten-one* rather than *eleven*), Asians tend to be able to absorb numbers and make calculations generally far more quickly than those in the West

What he writes has huge implications for presentations. It means that if you are talking to an audience that includes a good number of people from the West (particularly the US and GB), you should try to:

1. work very hard yourself to make it absolutely clear what you are saying, so that it is effortless for the audience to understand

2. be aware that your audience may not be used to concentrating for long periods and may thus have a short attention span

3. give the audience time to absorb and understand any numbers and statistics that you give them

4 PREPARING A SCRIPT

4.1 Decide what you want to say before you start preparing the slides

Most people prefer to spend more time creating slides than deciding what they are going to say about those slides. In fact, often they spend so much time preparing the slides that they have little or no time to think about the accompanying speech. This usually has disastrous consequences.

It makes far more sense to carefully plan (and practise) what you are going to say, before you prepare the slides. This means that you can avoid wasting time preparing unnecessary slides. your slides should reflect what you want to say. Your slides should not control what it is you will say.

4.2 Create an initial structure

Think of what questions your audience will want answered. Use these questions to help you to organize what to say in your presentation. Here is an example set of questions:

What is the current situation?

Why is this a problem?

What should be done?

Benefits of doing it.

Drawbacks / Costs of doing it.

What happens if it isn't done?

Alternatively, write down what you think are the most important / interesting aspects that you want to communicate to your audience. Try to limit the number of your important points (hereafter, key points) to about three or four, which is what most audiences can realistically remember. By not trying to cover everything but limiting yourself just to certain aspects, your presentation will have a clear focus. This does not mean that you only mention these key points and nothing else. Instead, it means that you mention the key points in your introduction and in your conclusions, and that you also give these key points the most space during the main body of the presentation.

A. Wallwork, *Presentations, Demos, and Training Sessions,*
Guides to Professional English, DOI 10.1007/978-1-4939-0644-4_4,
© Springer Science+Business Media New York 2014

4.3   Record yourself chatting about the main topics

Now record yourself talking about the main topics. Imagine you were just chatting to a friend. This will help you find an informal style, which is the style most audiences prefer. When you listen to your recording, you will understand:

- if the order you chose to present your points was the most logical and simplest to follow
- what words or phrases are difficult for you to say
- what examples you could give

You will also realise that talking in your normal voice (rather than a professional 'presentation' voice) is the best way to deliver a presentation. It makes you sound sincere and in personal contact with the audience.

4.4 Transcribe your recording and then check for naturalness and relevance

Transcribe what you have said. Make changes, but try to keep the colloquial style. Listen out for phrases that don't sound natural. Ask yourself: "Is this something that someone would say in a natural conversation?" If it isn't, change it.

Then classify each point in your presentation as follows:

- A: absolutely essential
- B: important
- C: include only if time

It might help you to do this if you imagine that the length of your allocated time for the presentation / demo was reduced, for example, from 60 min to 30 min. Decide which points:

- the audience might already know or not be interested in
- you have included simply because you think you SHOULD include them, because you think it is more professional to cover everything or because you think by putting them in you will make a good impression on your boss
- you could put in the handout as extra information without affecting the main logical argument of your presentation (the audience might prefer to read the details at their own leisure and at their own speed)
- you have included simply because you find them interesting, but they are in fact not particularly relevant
- could be grouped together under one category so that they could be covered together and more quickly

Would your presentation not succeed as a result? Or would it actually be clearer and more dynamic?

4.5   Write out your speech—or at least the critical parts

For the technical parts of the presentation, it may be enough to write notes. This is because these aspects will probably be the easiest for you to talk about, as you will be very familiar with them and will probably have all the correct English terminology that you need.

If you don't have the time and / or money to write a speech and have it revised, then try to make your English as perfect as possible in the following parts of the presentation:

1. in the introduction

2. while explaining the agenda

3. when making transitions from one series of slides to another series

4. when asking questions

5. in the conclusions

6. when calling for questions

These are the points when the audience will notice the mistakes the most and when they are forming their first and last impression of you, i.e. the impressions that will remain with them after the presentation.

In each minute of a presentation / demo you are likely use around 120–180 words, depending on how fast you speak and how much time the audience need to absorb the slides. Thus, if you write down exactly what you are going to say in the five key parts mentioned above, this will not require more than a couple of pages.

The reason for writing the script is NOT for you to then learn every word. Memorizing a script is not a good idea. You will not sound natural when you speak and you might panic if you forget your 'lines'. In any case, the slides themselves will help you to remember what to say.

Instead, writing a script will help you to define the content and the structure, and thus to decide:

- what the best structure is and thus the best order for your slides

- if certain slides can be cut

- find the moments in the presentation where audience interest might go down

- clarify where you need to make connections between slides

- verify if you are spending too much time on one point and not enough on another

- time how long the presentation will take

4.5 Write out your speech—or at least the critical parts (cont.)

A written script will also help you clarify the exact words that you will need to use. You will be able to:

- identify words that you may not be able to pronounce

- check that the sentences are not too long or complex for you to say naturally and for the audience to understand easily

- understand when an example would be useful for the audience

- delete redundancy and unnecessary repetition

- check if there are any terms that the audience might not understand

- think of how you could deliver your message in a more powerful or dynamic way

- check that any questions you ask (real or rhetorical) are constructed correctly e.g. *So why are we doing this?* (Rather than, *So why we are doing this?*)

Once you have written your script, you can then write the slides.

In addition, if you write a speech, then you can easily email it to an English speaking colleague to revise. Then you can be sure that at least the grammar and vocabulary will be correct. This is also an easy means to show your speech to a colleague (without forcing him / her to watch you performing)—this is a quick way to see if your presentation is clear and interesting.

4.6 Decide what style to adopt

Look at these two versions of an oral speech which give identical information. Which do you prefer? Which do you think the audience would prefer?

A

The main advantages of these techniques are a minimum or absent sample pre-treatment and a quick response; in fact due to the relative difficulty in the interpretation of the obtained mass spectra, the use of multivariate analysis by principal component analysis, and complete-linkage cluster analysis of mass spectral data, that is to say the relative abundance of peaks, was used as a tool for rapid comparison, differentiation and classification of the samples.

B

There are two main advantages of these techniques. First, the sample needs very little or no pre-treatment. Second, you get a quick response. Mass spectra are really hard to interpret. So we decided to use two types of analysis: principal component, and complete-linkage cluster. We did the analysis on the relative abundance of peaks. All this meant that we could compare, differentiate and classify the samples.

What problems do you think you would have if you had to say Version A aloud? And what problems would the audience have in understanding it?

Version A would be difficult to understand even if it were in a written form. And an audience at a presentation would find it hard to assimilate so much information at a single time. And for the presenter, it would be hard to breathe while saying such a long sentence (74 words!) without a pause.

The solution is to:

- split the sentence up into very short chunks (12 words maximum) that are easy for you to say and easy for the audience to understand
- use more verbs (the original contains only 4 verbs but around 20 nouns)
- use the active form and personal pronouns

Version B contains a series of short phrases. Short phrases do not mean that you express yourself in a simplistic way. You can give exactly the same information and keep all the technical terms that you need. The result is something that sounds natural and that the audience will enjoy listening to. If you talk like in the first version you risk alienating or confusing your audience.

4.7   Choose the right level of formality

The style of language you adopt in your presentation will have a huge impact on whether the audience will:

- want to listen to you and their level of enjoyment / interest

- find you approachable and thus someone they might like to collaborate with

There are essentially three levels of formality:

1. formal

2. neutral / relatively informal

3. very informal

Most audiences prefer presenters who deliver their presentation / demo in a relatively informal way. In English, this informality is achieved by using:

- personal pronouns (e.g. *I, we, you*)

- active forms rather than passive forms (e.g. *I found* rather than *it was found*)

- verbs instead of nouns

- concrete or specific nouns (e.g. *cars*) rather than technical or abstract nouns (e.g. *vehicular transportation*)

- short simple sentences rather than long complex ones

The secret in presentations is thus to be seen as being both authoritative and competent, but also as friendly and warm. The two are not incompatible—the authoritativeness comes from *what* you say, the friendliness from *how* you say it.

So when you finish writing your script, check that each sentence sounds like something that you might say to a colleague at lunch time. If it isn't, rephrase it in simpler terms so that the audience will feel that you are talking directly to them. This has big advantages for your English too. The simpler your sentences are, the less likely you are to make mistakes when saying them.

4.8 Only have one idea per sentence and repeat key words

Use the simplest English possible by using short phrases containing words that you find easy to say.

Each sentence should only contain one idea. This makes it easier for you to say and for the audience to understand.

Split up long sentences by deleting relative pronouns (*which, who, that*), and link words and phrases (e.g. *and, also, however, moreover, in addition, it is worth noting*)—such words are in bold in the 'Original' example below.

ORIGINAL	REVISED
The scenario is a typical wireless network, in **which** *there is a single base station in the middle and subscriber stations around it. We used a simulator in order to understand how the power saving mechanism influences the performance of the users* **and in addition** *to calculate what effect it has on the environment. It* **is also worth noting** *that testing can be classified in different ways depending on which part of the network is being tested and on how the testing is being carried out.*	*The scenario is a typical wireless network. There is a single base station in the middle and subscriber stations around it. We used a simulator to help us understand two factors. First, how the power saving mechanism influences how users perform. Second, the effect that power saving has on the environment. Another important aspect. [pause] Testing. [pause] Testing can be classified in different ways depending on which part of the network you are testing and on how you are doing the testing.*

Notice how in the revised version:

- the sentences are much shorter. This gives you natural pauses when you're speaking
- key words have been repeated in the place of pronouns (e.g. *power saving* instead of *it*). This helps the audience to follow you as they may not remember what *it* (or similarly *they, this, that* etc) refer to
- verbs are used in preference to nouns (e.g. *how users perform* instead of *the performance of the users*)
- *emphasis and drama can be created by very short phrases interspersed with pauses*
- *active forms are used instead of passive forms (final sentence)*

4.9 Break up long sentences that have parenthetical phrases into shorter sentences

Your audience needs to be able to understand you immediately—they are not reading, therefore they do not have an opportunity to re-hear what you have said.

So remove parenthetical phrases, as indicated in the 'short sentences' column below.

This will

* give emphasis and drama
* give you natural pauses when you're speaking

LONG SENTENCES	SHORT SENTENCES
The second important **thing** is that **testing** can be classified in different ways on the basis of the part of the system being tested **and** how testing is performed.	The second important **thing. Testing** can be classified in different ways on the basis of the part of the system being tested. **And** on how testing is performed.
Finally, in the main part of the presentation, **which** should be the most interesting part for you, we'll see how this process can be tailored to your needs.	In the main part of the presentation, we'll see how this process can be tailored to your needs. **This** should be the most interesting part for you.
This may mean that the problem is then ignored **even though** it might be a really important problem which could cost the company a lot of money.	This may mean that the problem is then ignored. **The thing is / But**, it might be a really important problem which could cost the company a lot of money.

4.10 Simplify sentences that are difficult to say

Create sentences that you find easy to say. Writing a script will help you to identify sentences, such as the one in the original version below, that are not easy or natural for you to say. So, read your script aloud, underline any phrases that are difficult to say, and then try to rewrite them until you find a form that is easy for you.

VERSION 1	VERSION 2
Most people speak at a speed of one hundred and twenty to two hundred words per minute, but the mind can absorb information at six hundred words per minute.	*a) Most people speak at a speed of **nearly** two hundred words per minute. **However**, the mind can absorb information at six hundred words per minute.*
	b) Most people speak at a speed of around two hundred words per minute. However, the mind can absorb information at six hundred words per minute – that is 400 words more than speaking.

Version 1 is difficult to say because it contains a lot of numbers plus a repetition of sounds (twenty *to two* hundred). Version 2a gives a more approximate number and splits the sentence into two parts. Version 2b states the same fact in a different way so that the audience will remember it better.

4.11   Be concise—only say things that add value

The more words you use

*   the more mistakes in English you will make!
*   the less time you have to give the audience important technical info

Here are some examples of sentences from the beginning of a presentation that could be deleted because they delay giving important information to the audience.

The title of my presentation is.

The product that I am going to present to you today is.

My presentation always begins with a question.

I have prepared some slides.

Here are some phrases that could be reduced considerably, as shown by the brackets:

Testing [can be considered an activity that] is time consuming

The main aim of this project, [as already shown in the previous slides,] is to find new methodologies for calculating stress levels. [In order to do this calculation,] we first designed.

4.12 Never delay key information: 1) topic 2) explanation and background

When you want to introduce a key topic, mention that topic immediately. Don't precede it with an explanation. So: 1) topic 2) explanation

If you make the audience wait to hear what the topic is while you are giving them preliminary information, they may stop listening to you. The topic in the example below is 'Template'.

ORIGINAL	REVISED
Finally, to help people to manage these types of documents by using some standards that help all the people in the company to work, a particular project has been started. It is an internal project called _'Template'. This could be_ used to store all the templates produced for reports, demos, minutes and presentations. Here is an example of a presentation template [shows example]. So everyone who does a presentation has to follow this template.	Now I want to move on to a new project called _'Template'. Imagine you want to prepare a new_ presentation for a client, or write out the minutes of a meeting you've just had or generally just produce a new document. Well, now you have templates to do this. Then you can store all the documents you produce on Template. Let me show you an example of a presentation template [shows example]. So whenever you want to prepare a new presentation you simply follow this template.

Note how in the revised version the presenter:

- immediately introduces the topic (the use of templates)

- gives the audience a concrete situation that they can immediately relate to

- uses a clear and logical structure

Another key aspect is that the presenter involves the audience by referring to them directly by using _you_. Using _you_ rather than an impersonal form is an example of an audience-oriented style. Think about how you feel when someone uses your name when speaking to you. You feel happy that the person has remembered your name. The same is true for an audience. They like to hear you address them personally. Obviously you can't always address them by name, but you can say _you_.

4.13 Reduce any introductory phrases when describing diagrams and examples

It is really tedious listening to someone who begins each slide or each bullet with 'Regarding x ....'

So if you are showing a new slide, instead of saying 'Regarding the prototype, here is the design' say:

> So here is the design of the prototype.

> Let's have a look at the prototype now.

The table shows some more examples of how you can reduce such redundancy.

REDUNDANT	CONCISE
Regarding the analysis of the samples, we analyzed them using...	We analyzed the samples using...
Regarding the design, it is very innovative.	The design is very innovative.
Here I present a panoramic view of the architecture.	This is the architecture.
Now you can see here an example of an interface.	Here is an interface.
We shall see two examples in the following slide.	So here are two examples.
In conclusion we can say...	Basically,...
In this picture I will show you a sample.	Here is a sample.
Another thing we wanted to do was...	We also wanted to...

4.14 Occasionally say what something cannot do rather than only focusing on what it can do

It is often useful to have a slide that says what something CANNOT or WILL NOT do—this may be as powerful as saying what it can do. A negation can have a positive effect. For example: *With our system you will not have the following problems:*

4.15 Have clear transitions from one part of the presentation to the next

You know the sequence of your slides and why they follow a particular structure. But your audience does not.

You need to help the audience follow your presentation. You cannot jump from one slide to the next at great speed. If the audience misses one particular point, they may lose the thread (i.e. the links, logical flow) of the rest of the presentation.

The way you move from one slide to another, and from topic to topic, is crucial. For the audience it should be like following a map, and you need to make it very clear to them whenever you make a turn. Also, at each turn it is helpful if you summarize for them what you have told them so far. Those in the audience who missed a previous turn, now have an opportunity to get back on the right road. This is different from a written document, where readers can, if necessary, just retrace their steps.

In a presentation, these moves or turns are called transitions.

A transition slide could be any of the following:

- simply a slide with a title on it
- a copy of the agenda slide but showing where you are up to—you can use grey to highlight what you have already covered and blue what you are going to cover now
- a mini new agenda of what you are going to do next
- a mini summary of what you have done so far

Before you move to the next section or group of slides:

1. pause for two seconds. This signals to the audience that you are going to say something important
2. look at the audience and give a quick summary of the most important things you have said so far. Repetition may seem boring to you because you know the subject so well, but it gives the audience a chance to check their understanding
3. explain how the next section relates to what you have just been talking about

This whole process should only take about 20 seconds, so it will not increase the length of your presentation unnecessarily.

4.15 Have clear transitions from one part of the presentation to the next (cont.)

A transition is also a good opportunity for:

- you to slow down or change the pace of the presentation
- the audience (and you) to relax a little—remember that the audience cannot assimilate vast quantities of information in quick succession
- you to regain the audience's attention by making them curious about what is coming next.

4.16 Be concise when making transitions

If you don't practice what to say when making transitions, you will probably improvise and say something like:

OK, that's all I wanted to say at this particular point about the infrastructure. What I would like to do next in this presentation is to take a brief look at the gizmo. This picture in this slide shows a gizmo. As you can see a gizmo is a ...

Instead of attracting the audience's attention, the above text is full of redundancy and adds no information. The audience is likely to go to sleep.

Try to make your transitions memorable.

OK, here's something that you may not know about a gizmo: blah blah blah. In fact you can see here that a gizmo is ...

4.17 Use a different transition each time you move on to the next point

Imagine you are talking to a client about the enhancements to one of your products that the client already has. Your aim is to get the client interested in buying the new version. When you introduce each new enhancement, it is a good idea to do this in a different way each time to create more variety and excitement.

OK, so we've looked at x, now let's look at y, which I personally think is probably the most useful enhancement.

OK, so the next enhancement has been requested by a lot of our customers, so I am really curious to hear your opinion on this.

The next enhancement should be quite quick for me to explain. Actually, it's one that we have been thinking about implementing but haven't made a final decision. It would be great if you could help us reach our decision.

Right, we have two more enhancements to look at.

So, we're at the last enhancement now. This is the one that our developers are really excited about.

4.18 Vary the grammatical forms that you use in your explanations and don't introduce each set of similar slides in the same way

During a demo or training session you often have to explain a series of new features, or changes to existing features. It can be boring for the audience if you constantly use the same grammatical forms or expressions each time you talk about a feature.

Below are two examples of how to introduce variation in the way you present features.

1) A speed dial has been added.

= There is now a speed dial.

= You'll probably be pleased to know that we have added a speed dial.

= Another useful feature is the speed dial.

= In the new version you will notice that there is a speed dial.

= Unlike the previous version, there is a speed dial.

2) X will allow you to do Y.

= With X you will be able to do Y.

= Having X means that you can do Y.

= So now you can do Y by using X.

4.19 Prepare your conclusions

Prepare your closing and know exactly what you are going to say / do. If you have no conclusions your audience may

• feel that you ran out of time and didn't actually finish the presentation

• feel that you are an unprofessional presenter who didn't even have the time to prepare the ending

• forget what the main points were

4.20 Consider not ending your presentation with a question and answer session

Many presentations end with a Q&A session (see Chapter 15). Experts however recommend finishing with something more memorable for the audience. For example:

- a brief summary of your key points (if you didn't make such a summary before the Q&A session)

- a further benefit to the audience of what you have proposed in your presentation

- some action points, i.e. what actions you would like them to take

- a very short anecdote that in some way encapsulates your message

- an amusing slide (e.g. a cartoon or photo) that sums up your message

Then remember to thank the audience for coming.

4.21 Revise your script to make the language more dynamic and effective

When you have finished the first or second draft of your speech, you can make it more effective if you follow the guidelines below:

USE VERBS RATHER THAN NOUNS

Using verbs rather than nouns (or verb + noun constructions) makes your sentences shorter, more dynamic and easier to understand for the audience.

X is meaningful for *an understanding* of Y. = X will help you *to understand* Y.

When you take into *consideration*... = When *you consider.*

This gives you the *possibility* to do X = This means you *can* do X. / This *enables* you to do X.

AVOID ABSTRACT NOUNS

Abstract nouns such as *situation, activities, operations, parameters, issues* are more difficult to visualize than concrete nouns and thus more difficult to remember. Often they can simply be deleted.

Our *research and development* ~~activity~~ *focused on.*

If you find that your speech is full of words that end in *-ability, -acy, -age, -ance, -ation, -ence, -ism, -ity, -ment, -ness, -ship,* you probably need to think about deleting some of them or finding concrete alternatives or examples.

AVOID GENERIC QUANTITIES AND UNSPECIFIC ADJECTIVES

Replace generic quantities such as *some, a certain quantity, a good number of* with a precise number.

I am going to give you a few examples = three examples

We have found some interesting solutions to this problem = four interesting solutions

Audiences like numbers:

- they make us more attentive because we start counting and we have a sense that we will be guided

- they give the information a more absorbable structure and thus help us to remember it better

4.21 Revise your script to make the language more dynamic and effective (cont.)

AVOID VAGUE EXPRESSIONS OR HIGHLY SUBJECTIVE ADJECTIVES

Statements such as *I think these new features will be very useful* are completely meaningless. You need to describe the features in such a way that it will be clear to the audience that they are useful. Try to say something more concrete. *These features should speed up the time it takes you to do x, and should also keep your customers happy because...*

Avoid vague phrases connected with measurement: *to a certain extent, more or less, a good number of*

OCCASIONALLY USE EMOTIVE ADJECTIVES

If you tell the audience you were 'excited' about something, then they are more likely to become excited too, or at least more receptive to what you are going to tell them. Good adjectives to use, for example in descriptions of diagrams or when giving results, are: *exciting, great, amazing, beautiful, incredible,* and also *unexpected* and *surprising* when used in a positive sense.

4.22   Use your script to write notes to accompany your slides

Most presentation software allows you to write notes for each slide. On the basis of your script you can write down what you want to say for each slide in note form. You can then print your slides with the accompanying notes and have these next to you when you do your presentation. It is best to print several slides on one page, then you don't need to keep turning the pages. Having these notes with you will give you confidence, because you know that you can consult them if you forget what to say, or forget where you are in your presentation. Alternatively upload your presentation notes onto your tablet or mobile phone (see 9.18).

Also, you can practice (see Chapter 16) your presentation using these notes.

4.23   Use your speech for future presentations

Having a written speech will also help you in future presentations. You may need to do the same presentation several times even several months or years in the future. You may be able to use exactly the same presentation, so practicing for it will be much easier if you already have a script. After each presentation it is worth going through the script to modify it and improve it in the light of the audience's reaction and questions. You will see where you need to add things and where to cut parts that weren't necessary, that the audience didn't understand, or which you found difficult to explain.

Even if you do a completely different presentation in the future, the way you introduce yourself and make your concluding remarks is likely to be the same—so this part of your script is worth keeping.

4.24 An example showing the advantages of writing a script

The two versions below highlight the advantages of writing out your script. The version on the left is the version that the presenter, Luigi, wrote himself. It contains many mistakes (highlighted in bold). He then sent this version to a native speaking English colleague, who produced the corrected version on the right.

UNCORRECTED VERSION: MANY MISTAKES	CORRECTED VERSION: NO MISTAKES
Hi to all, thank you to be here today; some of you just know me - for the others: my name is Luigi, I am working at ABC in Boston since the last 9 years, in various teams. I've started developing video games for the under 10s market and I've then been moved to the core XTC video team and after that to the WEB project team, where I am senior developer. Thanks to the experiences we've had with both the XTC and WEB projects, we have tried to prepare these presentations so that we can give the other work groups in the company details of how to exploit both these technologies and to propose you some design patterns that should simplify your everyday life while you're ...	*Hi everyone, thank you all for coming here today. For those of you who don't know me, my name is Luigi. I've been working at ABC in Boston for the last 9 years. I've been lucky to work in a wide variety of teams. I started off by developing video games for the under 10s market. I was then moved to the core XTC video team and now I am in the WEB project team, where I am a senior developer. So, I hope that my experience both in the XTC and WEB projects mean that I am in a good position to give you some useful details of how to exploit both these technologies. Also, the idea is to show you some design patterns that should really simplify your everyday life when you're ...*

Note how the corrected version has eliminated the many grammar mistakes (highlighted in bold):

- *thank you* is not followed by the infinitive (so *thank you for being here* not *to be here*);
- present perfect continuous (*I've been working*) for something that began in the past and is still true now
- before job positions we normally put the indefinite article
- simple past for finished actions (*I started off*)
- use of articles: *I am a senior developer*, not *I am senior developer*

4.24 An example showing the advantages of writing a script (cont.)

It should also eliminate other mistakes, for example:

- expressions such as *Hi to all* (*Hi all* would be in an email, but when speaking you say *Hi everyone*)
- the difference between *for* and *since,* and between *just* and *already*

By reading what you are going to say you can also make other improvements that would be hard to make if you only practised orally. For example, making your script more:

- dynamic by introducing words such as *lucky* (e.g. *I've been lucky to work in a wide variety of teams*), and other emotive words such as *great, pleasure, happy.*
- motivating to the audience by using *you* rather than impersonal expressions such as *other work groups in the company*

These kinds of corrections and changes are difficult to make if you don't produce a written script to work on.

5 PREPARING YOUR SLIDES

5.1 Ensure each slide has a purpose

If you have prepared a script following the guidelines in Chapter 4, you will now be reasonably familiar with the content of your presentation. This is the moment to decide which slides are really needed.

Every slide should have a purpose and its purpose must be clear not just to you but also to the audience. A slide is needed when it does one or more of the following:

- makes an explanation less complicated and quicker

- helps people to visualize and recall something better

- makes something abstract become more concrete

- attracts attention or entertains the audience (but only in a way that is relevant to your topic)

If a potential slide does not do any of the above, then you probably do not need to create it. You do not need a slide for every point you make. Some points you can simply tell the audience or alternatively write them on the whiteboard.

A. Wallwork, *Presentations, Demos, and Training Sessions,*
Guides to Professional English, DOI 10.1007/978-1-4939-0644-4_5,
© Springer Science+Business Media New York 2014

5.2 Decide which points do not need an associated slide

To help you decide where slides are absolutely necessary, tell another colleague all about the topic of your presentation and use a whiteboard where necessary. You will probably notice that you can say a lot of things without any slides at all, but simply by talking with the occasional aid of a whiteboard.

Your presentation will be much more dynamic, and more varied for the audience, if you have periods when you simply talk without any slides on the screen.

Because of the type of documentation you write in your working lives and because of the presentations you regularly see from your colleagues, your natural instinct is to present all information in the form of headings, subheadings, bullets and often unnecessarily complex diagrams (that have been pasted from technical docs).

However, audiences do not appreciate seeing one slide after another of bullets, texts and diagrams.

Now go back to your written version and choose which points don't really need to be in a slide format but could either be:

* spoken, i.e. a slide isn't necessary, the info can simply be imparted by voice

* put on the whiteboard, i.e. a slide is necessary, but the info could equally well be shown on the whiteboard

5.3 Limit yourself to one idea per slide

Each slide should only have one main idea. Thus any bullets, data, or graphics on the slide should be in support of this main idea. You can check how many ideas there are in your slide by trying to give it a title. If a title doesn't come quickly to mind, it may mean you have covered too many points and thus that you need to divide up these points into further slides.

The moment to give detail is when you are talking through the slide. There shouldn't be too much text / detail within the slide itself.

5.4 Choose and effective title slide

Try to make the title of your presentation as specific and interesting as possible. A series of generic abstract words is not likely to inspire people to come to your presentation.

You can find some very good examples here (including complete presentations):

http://www.topdesignmag.com/25-fantastic-powerpoint-presentations-for-your-inspiration/

The title slide could also include your name.

5.5 Decide on a system of capitalization for titles and subtitles

In your title slide you may wish to use initial capital letters for the key words.

Five Ways to Move forward

Alternatively you may think it would look better for the first letter of every word to be capitalized:

Five Ways To Move Forward

Or you could just have initial capitals for the first word:

Five ways to move forward

However, proper names, months, days, countries etc must always have initial capitalization:

Five ways to move forward in Europe

The most important thing is to be consistent throughout the presentation. Below are the various approaches in more detail:

APPROACH 1

Capitalize the first word. Then capitalize all other words except:

articles (*the, a, an*)

conjunctions (*and, but, or* etc)

prepositions (*at, in, on* etc)

to in infinitives

Examples

The Way Forward in Computer Science

New Ways to Get to Heaven

APPROACH 2

In the title slide (i.e. the first slide) capitalize the first word and all the important words (but not articles, conjunctions, prepositions and *to*).

Example

The Way Forward in Computer Science

Simone Garfunkel
26 October 2020

5.5 Decide on a system of capitalization for titles and subtitles (cont.)

Then, only capitalize the first word in all other titles (this is the easiest option).

Examples

The way forward in computer science

New ways to get to heaven

APPROACH 3

Do not capitalize any words at all. However, more traditional members of your audience may find this strange. If you adopt this method, it is vital that it stands out as a title (use bold or a different color).

5.6 Minimize the number of items on your agenda slide

Have one slide that tells the audience what you are going to do during the presentation. Then you probably need another slide to remind you to tell the audience:

- how long the presentation will take

- if there will be breaks

- when the audience can ask questions and if they can interrupt if they can't follow

- whether there is a handout (3.6)

- that the presentation can be found on your website

Ensure that your agenda slide is not full of bullet points and text.

The agenda below is for a course on writing emails. Imagine how the audience might react.

Course Agenda

- [0] Introduction and objectives

- [1] Subject lines

- [2] Initial salutations

  formal emails
  informal emails

- [3] Final salutations

  formal emails
  informal emails

- [4] Main text

- [5] Replies

  to known recipients
  to unknown recipients

- [6] Automatic translation software

  the main advantages and plus points
  the main disadvantages and minus points

- [7] Practice session

The problem is that the audience will feel that there is an incredible amount of ground to cover, and thus the demo / presentation is going to be long and / or boring. This will demotivate them right from the beginning and they will be put in a negative frame of mind. Instead, you can:

5.6 Minimize the number of items on your agenda slide (cont.)

- have fewer points - in the example above there are eight points. The first point ([0] is *Introduction and objectives.* Given that most standard presentations always begin with an introduction and objectives, it is not necessary to have this information written on the slide. Of course, if you feel it is absolutely necessary, you can still tell the audience that you will begin with an introduction and objectives. Points [2] and [3] seem to be about very similar things, and so could be combined into one bullet

- remove all second level bullets. In the agenda above, the second level bullets provide more details on the first level. You can mention these details to the audience as you talk them through the agenda, but they do not need to appear on the slide

- remove any words that are not strictly necessary (see Chapter 6) – for example the word *emails* in [2] and [3]

- minimize the number of words, e.g. in [6] you could simply write *Automatic translation software: pros and cons*

- simplify the look. In the agenda above why are there bullets and numbers? Why do the numbers begin with [0]? The presenter may know the answers, but it is unlikely that they are relevant for the audience

- get the audience excited. From the agenda it seems that in most of the course the participants will be passive. They only get to do hands-on stuff at the end: *[8] practice session.* Particularly if you are trainer you need to have a constant mix of practise and theory, not just all theory and then practise.

Below is a new version of the original agenda, revised on the basis of the points above.

Agenda

1 Subject lines

2. Initial and final salutations

Practice session

3 Main text

4 Replies

Practice session

5 Automatic translation software

Practice session

5.7 Don't create a slide without thinking what you are going to say about it

When you create a slide think about

- why you are going to tell the audience about x, and which part of x they need to know
- an example of x
- how x will benefit them in some way
- what will happen if they don't do x

Then think about

- the best way to present x on the slide
- the best way to talk about x
- the kinds of questions the audience might ask about x, and have the answers already contained in the slide (if you don't have the space, then give the answers verbally)

5.8 Avoid too much text and complete sentences

If you fill your slides with text, you are:

• likely to read out the text word for word karaoke style

• encouraging your audience simply to read and not to listen to what you say. At this point you could simply email the presentation to your audience.

By simplifying and cutting, you will have much cleaner slides. The audience will then spend more time listening to you, and less time reading your slides.

Assuming your audience all understand English quite well, if you write complete sentences in your slides:

• your slide will be full of text and to accommodate this text the font may be too small for the audience to read clearly

• your audience will read the text on the slide rather than focus on you; also they will not all read at the same speed

• when you comment on the slide if will be difficult for you to avoid repeating word for word what is on your slide

Moreover, if you have a lot of text on your slides but you say something very different from the text, then the audience has to take in two different sets of information – one written, the other verbal – at the same time. The human brain is not equipped to simultaneously read some information and to listen to something different.

So the solution is to do one of the following:

• cut the slide completely and simply talk

• reduce the text to three or four short bullet points which the audience can absorb immediately. Then expand on one or more of these bullets

• give the audience a few seconds to absorb the text, and then blank the screen and start talking

Otherwise there will be two presenters - you and your text - and you will both be competing for the audience's attention.

However, complete sentences can occasionally be used to emphasize a particular point, explain a difficult point, or give a quotation or definition.

5.9 Help audiences with low level English

Some audiences, however, appreciate complete sentences. They enable attendees with a low level of English to:

- follow your slides, even if they can't follow what you say

- better understand your pronunciation if they can also see the written forms of the key words that your are using

- take notes

- memorize what you have said if they have a better visual memory than auditory memory

Two possible solutions for dealing with an audience with mixed levels of English are:

1. have slides with complete sentences but keep them as short as possible, removing all redundancy (see Chapter 6) and removing articles (*the, a/ an*). When you show these slides, give the audience up to 5-6 seconds to read them. Then, make general comments without reading the text. This allows the audience to absorb the information on the slide and then they can concentrate on what you are saying

2. have short bulleted sentences. In addition, prepare photocopies of the same slides but with full text (3.6-3.8).

5.10   Be careful how you use numerical examples

If you use numerical examples, make sure the numbers appear on the slide as it is very difficult for a non-native audience to immediately translate numbers and then be able to follow the example.

For instance if you say *revenues have not been steady. This is illustrated by the fact that revenues went up by twenty three point four five per cent in the first quarter but only zero point one seven six in the second quarter* the audience may get lost when listening to the numbers. Instead, if you show the numbers on the slide, it will be easier for the audience.

REVENUES

Last year

Total revenues: + 12.24%

This year:

Q1: + **23.45**%
Q2: + **0.17**%
Q3: + 2.45%

In the example above, the numbers that the presenter mentions are highlighted in bold. Other solutions for highlighting numbers are to circle them or to put them in a different color.

5.11   Explain processes through well chosen slide titles

When the main purpose of your presentation is to explain a process or how a piece of equipment works, it is a good idea to use your slide titles to explain each step in the process. Here are titles of a sequence of six slides from an engineering demo. Each slide simply has a title and then a diagram or picture, which the presenter then explains.

[Title slide] 3D Laser milling modeling: the effect of the plasma plume

Laser milling: a process well suited for mold manufacturing

Laser milling centers consist of various sub-systems

The laser beam is controlled by a laser beam deflection unit

A valid estimation of the material removal rate is required

Many parameters affect the material removal rate

5.12 Illustrate part of a process rather than the entire process

When illustrating a process, there is generally no need (or time) to show the whole process. Instead, just show the part of the process that the audience really needs to know or the part that is the most interesting. The entire process can be explained in a separate written document which you can then email to the audience.

If you show the whole process, this may confuse the audience and make you lose your focus.

Ignore any pre-existing graphics that you have of the full process, and create your slide from scratch. This does not have to be a laborious process, because you only need to highlight the essential.

If something is quite complex, then break it up into manageable steps over two or three slides - but occasionally go back one slide or two, to highlight the various connections to the audience.

To satisfy any potential criticism that you have been superficial, you can tell the audience where to find the full graph.

5.13 Ensure that everything your write on your slides is 100% grammatically correct

If you make mistakes in your English when you talk, the majority of your audience will probably not care or not even notice. However, they may notice written mistakes. Don't be creative with your English. Only write what you know is correct. Generally speaking, the shorter the sentence, the less likely you are to make a mistake.

However, the less text you have, the more evident any grammar or spelling mistakes are. These final slides from three different presentations (all real) did not make a good final impression:

End

Tanks!

Any question?

The presenters should have written 'The end', 'Thanks' and 'Any questions?'

If there are mistakes on your slides, you may be unable to see them - so it is always wise to get a colleague to check them.

The only time you are likely to see the mistakes is, ironically, while you are actually giving the presentation! And if you see a mistake, this is likely to undermine your confidence.

5.14   Check the spelling and use of capital letters

When you become very familiar with your slides it becomes almost impossible for you to notice spelling mistakes. It is also pssobile to udnresnatd cmpolteely mssiplet wrods and snteecnes. So this means you may not see the mistakes.

PowerPoint does not always manage to highlight incorrect spellings. To check the spelling of your presentation you need to convert the text into your word processing program (e.g. Word, NeoOffice). Microsoft Word highlights words that it thinks are not spelled correctly with a red underline. However, given that you probably use a lot of technical words, these too may appear with a red underline because they are not in Word's internal dictionary. It is easy just to ignore these words hoping (or presuming) that you have spelt them correctly. But there is a good chance that at least one of these words will not be spelt correctly. It is a good idea to check on Google or with an online technical dictionary whether the spelling is correct or not.

Some of your misspellings of normal words may not be highlighted because they are words that really exist. Some examples:

chose vs *choice, fell* vs *felt, form* vs *from, found* vs *founded* vs *funded, led* vs *leaded, lose* vs *loose, than* vs *then, through* vs *trough, with* vs *whit, which* vs *witch*

Generally speaking, the following types of words always require their initial letter to be capitalized:

- days of the week and months (Thursday, January)
- names of people and places (Smith, London)
- names of companies (Apple, Google, Camper)

5.15 Prepare a sequence of identical copies of your last slide

Typically if you hit the advance button while showing your last slide you will drop out of the presentation program. This then means the audience will see the smaller window of the presentation and your desktop - this does not look very professional. Duplicate two or three copies of your last speaking slide so that if you accidentally advance one too many times at the end of your presentation, the slide looks like it has not changed.

After these slides, you should include some slides that answer questions that you expect to be asked. These slides will be useful during Q&A sessions after the presentation.

5.16 Modify your script on the basis of the slides

You have now created your slides. The next stage is to modify your script so that it takes into account exactly what you will say about each slide.

5.17 Modify an existing presentation that someone else has created

Occasionally, you may have to use a presentation that a colleague has produced. Before you use it, first carefully remove any signs that this presentation has been given to a previous audience (e.g. the date of the previous presentation, references to names of companies, trade fairs, conferences etc).

Tailor the presentation to the needs of the new audience by:

- revisiting the agenda
- changing the slides themselves
- changing the order of the slides
- deleting and adding slides
- deciding to use the whiteboard rather than using particular slides
- injecting your own personal approach and priorities

If you don't tailor the slides, the presentation will always feel like someone else's, so you won't sound confident or convincing.

Remove any slides that you think you will not use, otherwise you will have to fast forward during the presentation. This will give the audience the impression that either the presentation was not specifically created for them or that you have suddenly realised that you have run out of time.

Moral of the story: Make the presentation your own.

6 HOW TO REDUCE THE AMOUNT OF TEXT AND NUMBER OF CHARACTERS

6.1 Remember that text on a slide differs from normal text

A slide differs from normal written text because it:

- doesn't have to respect the normal rules of grammar

- can contain abbreviations (e.g. *info*), acronyms (*asap*) and contractions (*we're*)

- can present info concisely by removing non-key words

6.2 Don't repeat the title of the slide within the main part of the slide

In the example below, the title of the slide has been repeated almost word for word in the introductory sentence to the bullets. Also 'remove redundant' is repeated three times.

HOW TO FREE UP SPACE ON YOUR SLIDES

The following are ways to free up space on your slide:

- remove redundant info

- remove redundant words

- remove redundant graphics

Here is a more concise version:

HOW TO FREE UP SPACE ON YOUR SLIDES

Remove redundant

- info

- words

- graphics

A. Wallwork, *Presentations, Demos, and Training Sessions,*
Guides to Professional English, DOI 10.1007/978-1-4939-0644-4_6,
© Springer Science+Business Media New York 2014

6.3   Choose the shortest forms possible

Use the shortest words and shortest phrases possible. Here are some examples:

regarding = on; however = but; furthermore = also; consequently = so; necessary = needed

We needed to make a comparison of x and y. = We needed to compare x and y.

There is a possibility that X will fail. = X may fail.

Evaluating the component = Evaluating components

The user decides his/her settings = Users decide their settings

The activity of testing is a laborious process = Testing is laborious

No need for the following: = No need for:

Various methods can be used to solve this problem such as: = Solutions:

The main requirements are as follows: = Main requirements:

6.4   Use only well-known acronyms, abbreviations, contractions and symbols

The following are examples of well-known and accepted acronyms and abbreviations:

as soon as possible (asap); *to be confirmed* (tbc); *for example* (e.g. or e.g.), *that is to say* (i.e. or ie); *info* (information); *against* (vs); *research and development* (R&D); *and, also, in addition etc* (& or +); *this leads to, consequently* (> or =)

To save character space, you can remove periods (.) from both acronyms (USA, UK, FIFO) and abbreviations (NB, eg, ie).

Don't use abbreviations, acronyms, and symbols unless they are well known. If you explain a new acronym in Slide 2, by Slide 3 the audience will already have forgotten what it means. It is much easier for them to see the full words.

6.5   Reduce the number of zeros

When writing documents, numbers from one to nine are generally written as words (nine) rather than Figs. (9). This is simply for readability, since the numbers may be initially confused for words (e.g. I vs 1). However this rule can be ignored if you have a problem with space.

You can reduce the number of zeros in long numbers by using standard abbreviations.

10,000 = 10 K

10,000,000 = 10 M

NB in Anglo countries a comma is used in whole numbers (80,000) and a decimal point in fractions (0.08).

6.6   Use the plural form of nouns to save space

Using the plural form of nouns rather than the singular can save up to six characters. For example:

Evaluating *the* component.	(25 characters)
Evaluating *a* component.	(23 characters)
Evaluating components.	(21 characters)
The user decides his / her settings.	(34 characters)
Users decide their settings.	(28 characters)

6.7   Cut brackets containing text

Brackets tend to contain examples, definitions, or statistics.

Natural fibers (wool, cotton etc)

ISO (International Organization for Standardization) approval

In the examples above, the parts in brackets can be removed from the slide. Instead, while you are explaining the slide you can say:

We analyzed some natural fibers such as wool and cotton.

Our device has been approved by the International Organization for Standardization.

By deleting the parts in brackets, you will thus have extra information to add when you comment on your slide, i.e. you will not need to read out your slide karaoke style!

6.8   Use verbs and modals rather than nouns

Verb constructions occupy less space than noun constructions, as highlighted in these examples:

We needed to *make a comparison* of x and y. = We needed to *compare* x and y.

There is a possibility that X will fail. = X *may* fail.

6.9   Remove *the, this, our*

If you are very short of space, you can remove certain words.

[The] main features [are]:

[This] system has five new features.

[Our] results prove that:

6.10   Don't put text in your slides to say what you will do or have done during your presentation

In an Outline there is no need to write:

I will discuss the following ...

Likewise on the Conclusions slide do not write:

I have presented a strategy for ...

In such cases, you simply need to say those phrases.

7 USING BULLETS

7.1 Choose the most appropriate type of bullet

Always use the standard bullet (•) unless the items:

- need to be numbered to show the order or chronology in which something is done

- are in a list of things that were scheduled to be done and have been done. In this case you can use a tick (√).

7.2 Limit yourself to six bullets per slide

When you are giving lists keep them short. Six bullets is generally more than enough. And you only need to talk about a couple of them (e.g. the top two).

An exception is when you are not going to talk about any of the bullets but your aim is simply to show that, for example, your product / service has a lot of features, or that your company has a lot of important clients. In such cases you can simply say *here is a list of the main features* or *here are some of our most important clients*, pause for a second, and then move to the next slide.

A. Wallwork, *Presentations, Demos, and Training Sessions,*
Guides to Professional English, DOI 10.1007/978-1-4939-0644-4_7,
© Springer Science+Business Media New York 2014

7.3 Keep to a maximum of two levels of bullets

The 'original' slide below has three levels of bullets, which generally leads to messy slides.

ORIGINAL	REVISED
GOALS	OPTIMIZATION GOALS
➢ Different optimization goals:	➢ Save storage
○ Save storage	➢ Save CPU utilization with multiple applications
○ Save CPU utilization	
■ Only if multiple applications are being run together	

As you can see from the revised version, you can reduce the bullets to one level by:

- Changing the title of the slide from *Goals* to *Optimization Goals*

- Incorporating the third level into the second level (*Save CPU use for multiple applications*). Alternatively you could delete the third level and simply give this information verbally.

7.4 Do not use a bullet for every line in your text

The default settings of PowerPoint and other applications encourage you to use a bullet before every line of text.

Note how the bullets in the original version below have been misused in this slide from a presentation on detecting faults in a magnet motor.

ORIGINAL	REVISED
MODELING FAULT CONDITIONS	MODELING FAULT CONDITIONS
➢ Two main faults are investigated:	➢Two main faults are investigated:
➢ Open phase. In this case the current sensor in each phase.	➢ Open phase. In this case the current sensor in each phase.
➢ Shorted turns. In this case a percentage of the turns of the winding is shortened.	➢ Shorted turns. In this case a percentage of the turns of the winding is shortened.
➢ Under these conditions the faulty ...	Under these conditions the faulty ...

The first line (*Two main faults ...*) introduces a list of two items. So only the second and third lines need bullets. The fourth line is not a *fault*.

7.5 Choose the best order for the bullets

The normal practice is to order the bullets in terms of which ones you will be commenting on. Given that there is generally no need to comment on all the bullets in a list, it is best to put the ones you intend to talk about at the top of the list.

Sometimes you may have a list of bullets and you intend to make one general overall comment, without commenting on any of them individually. In such cases it is best to put them in alphabetical order to highlight that they are not in order of importance. Alternatively, you can say: *By the way these bullets are in no particular order.*

7.6 Use one-at-a-time bullet animation only if absolutely necessary

Presentation applications allow you to introduce items in a list one at a time. This can be useful if it is crucial to delay information, for example when giving your conclusions in order to get the audience to focus on one conclusion at a time.

Otherwise, show all the items at once and give the audience three to five seconds to absorb them before you start talking. This means that

- you don't have to keep hitting the mouse to introduce the next item. Your hands are thus free and you can move away from the laptop and keep your eyes focused on the audience

- the audience doesn't have to constantly keep changing where they are looking (you or your slides), and they are not waiting for the next item to appear. They can do all their reading at once

- you won't inadvertently introduce two items at the same time (and thus lose the whole point of delaying the information)

7.7 Make good use of the phrase that introduces the bullets

To save space, don't repeat the first words in a series of bullets.

ORIGINAL	REVISED
The advantages of using this system are:	Advantages for users:
➢ *it will enable users to* limit the time needed online	➢ limits online time
➢ *it will help users to* find the data they need	➢ finds relevant data
➢ *it will permit users to* get more accurate results	➢ produces more accurate results

In the 'original' example above, the first four words on each bullet (*enable, help* and *permit*) mean the same in this context.

7.8 Use verbs not nouns

Where possible, use verbs both in the introductory sentence and in the bullets themselves. Using verbs, rather than nouns, reduces the number of words you need.

NOUNS	VERBS
Testing is the activity of	Testing involves
➤ the observation and recording results	➤ observing and recording results
➤ the evaluation of the component	➤ evaluating the component

7.9 . Be grammatically consistent

Make sure the first word in each bullet is grammatically the same:

- an infinitive (e.g. *work / to work*)
- an *-ing* form (e.g. *working*)
- a verb (e.g. *works / will work*)
- a noun (e.g. *employee*)
- an adjective or past participle (e.g. *good, better, improved*)

GOOD: ALL VERBS	GOOD: ALL ADJECTIVES / PAST PARTICIPLES	BAD: 1 NOUN; 2 VERB; 3 ADJECTIVE
Advantages for users:	Advantages for users:	Advantages for users:
1. *reduces* costs	1. *reduced* costs	1. *reduction* of costs
2. *finds* relevant data	2. *relevant* data	2. *finds* relevant data
3. *produces* accurate results	3. *accurate* results	3. *accurate* results

The grammar in the slide in the second and third columns below may initially look correct, but it isn't.

GOOD: ALL -ING FORM	BAD: UNGRAMMATICAL	BAD: UNGRAMMATICAL
A Java infrastructure for	A Java infrastructure for	A Java infrastructure for
1. process*ing* MPEG-7 features	1. MPEG-7 features processing	1. MPEG-7 features processing
2. manag*ing* XML database	2. XML database managing	2. XML database management
3. exploit*ing* algorithms ontology	3. algorithms ontology exploiting	3. algorithms ontology exploitation
4. integrat*ing* functions	4. functions integrating	4. functions integration

In the second column, the final word in each bullet ends in *-ing*, but unfortunately they are not all the same grammatical form. *Processing* can be a verb or a noun, but the other three (*managing, exploiting, integrating*) can only be verbs and cannot be in this position in a phrase. In the third column, there is a series of noun+noun+noun constructions, which is difficult for the audience to understand quickly and in bullet points 1, 3 and 4 this construction is not grammatically correct. The best solution is to use verbs, as in the first column.

7.10   Minimize punctuation in bullets

There is no general agreement on how to punctuate bullets. The simplest solution is to use no punctuation at all, and begin each bullet either with a lower case letter or with an upper case letter. The phrase that introduces the bullets is generally followed by a colon (:) but this colon is not obligatory.

Advantages for R&D department:

- limited lab time
- relevant data
- more accurate results

However, some people prefer to punctuate:

Advantages for R&D department:

- limited lab time;
- relevant data;
- more accurate results.

The problem is that the punctuation marks make the slide look less clean and the extra character space required by the punctuation may make the bullet point run onto the next line.

In any case, ensure you use the same style consistently.

8 USING AND COMMENTING ON DIAGRAMS AND PROCESSES

8.1 Simplify everything

Given that tables and graphs are difficult to interpret quickly, decide if it would be possible to present the same information in a much clearer way.

A sequence of related tables over several slides means that the audience have to remember what was in the previous tables. The best solution is to have all the information on one slide. You can only do this by significantly reducing the amount of information and having a maximum of two adjacent figures.

8.2 Only include visuals that you intend to talk about

If you present a slide full of information, you yourself know what is important and where to focus your eyes, but the audience doesn't.

Only show graphs, charts, tables, and diagrams that you will actually talk about. If you don't need to talk about them, you could probably cut them.

8.3 Avoid visuals that force you to look at the screen

A key quality of good presenters is that they spend 95 % of their time looking at the audience. They minimize the moments when they need to look behind to see what is on the screen.

If you talk while looking at the screen you lose audience attention and also your voice is much more difficult to hear.

If your visuals are clear you shouldn't need to look at the screen or point. If you need to point, it may mean that you need to simplify what is on your slide. Simplification is obviously a benefit for the audience but also for you because it means that you will not get lost or confused by having to give complicated explanations.

The problem with pointing with your hands / fingers, your cursor, or using a laser pointer is that it may be clear to you where you are pointing but it rarely is for the audience. It also means that you will have to turn your back on the audience for several seconds. This can be very distracting for the audience.

A. Wallwork, *Presentations, Demos, and Training Sessions*,
Guides to Professional English, DOI 10.1007/978-1-4939-0644-4_8,
© Springer Science+Business Media New York 2014

8.4 Use visuals to help your audience understand

We tend to enjoy the creative graphical side of preparing a presentation but think less about the actual utility for the audience of what we have created. The aim of visuals is to help your audience to understand, but often they confuse the audience.

To avoid confusion, experts recommend:

TYPE OF GRAPH OR CHART	USEFUL FOR	MAX. NO. ELEMENTS
Pie	percentages	3-5 slices
Bar charts (horizontal), columns (vertical)	comparisons, correlations, rankings	5-7 bars / columns
Graphs	showing changes over time. Scatter graphs give clear overview of how data are scattered	1-2 lines
Tables	comparing small amounts of information	3 columns and 3 rows
Cartoons	clarifying all kinds of graphs and charts	1-2

In addition, you should:

• minimize the amount of information contained

• include labels and legends / keys, and locate them as close as possible to the data points they refer to

• ensure that labels are horizontal, otherwise the audience will find them difficult or impossible to read

• explain what the axes represent and why you chose them

• present comparative information in columns not in rows

You can also use visuals to:

• get audience attention

• inject humor

• vary the pace of the presentation

8.5 Make sure your slide can be read by the audience in the back row

Audiences will not be pleased if you say *I know that this is too small for you to read but ...* If a table or graph is too small or too detailed, it can be distracting and confusing. One solution is to enlarge just one part of it, i.e. the key element you want your audience to understand. If showing the whole table is essential for your purposes, you can show it all in one slide. Then in the next slide show a reduced version but highlighting the interesting part through color, circles or enlargement.

8.6 Choose fonts, characters, and sizes with care

It may be tempting to use lots of formatting because it makes slide preparation seem more creative. However, your text will be easier to read if you limit underlining, italics, shading, and other forms of formatting to the minimum.

The major organizations on the Internet (e.g. Google, Firefox, Amazon, YouTube) use Arial, or a similar font. Research has shown that if you use an easy to read font such as Arial or Helvetica, people are more likely to be persuaded about what you are saying.

Comic sans gives the idea of fun and children, and is thus probably not appropriate in a presentation. Presenters sometimes choose it because they think that by doing so they automatically give their presentation a fun element - but it is actually more difficult to read and does not look very professional. Times is possibly the most common font used for writing documents, but it is more difficult to read than fonts like Arial.

If you use a font size smaller than 28 points, the audience may not be able to read your slide. Use 40 points for titles.

Avoid writing a complete sentence in capital letters. Signs in airports, highways, and metropolitans are all in lower case letters. Why? Because capital letters are much more difficult to understand.

8.7 Limit the use of animations

Some features of presentation software often seem to be used solely to impress the audience. Animations are occasionally useful, but they:

- can typically and inexplicably go wrong during the presentation itself
- can be distracting and annoying for the audience
- tend to be used to explain complicated processes. It may be better to just simplify the process – the audience doesn't need to see or understand every step in the process (see 5.12).

8.8 Use color to facilitate audience understanding

Only use color to help the audience understand your visuals, not simply to make them look nice. Be consistent with color; use the same color for the same purpose throughout the presentation.

Website designers know that the background of a website can have a significant effect on whether a surfer is likely to stay and look and possibly buy. This implies that the background color of your slides may also affect how willing the audience will be to spend time looking at them. The experts suggest using dark text such as blue or black on a medium-light, but not bright, background, or light colors on a medium-dark background (e.g. yellow on blue). Dark colors on a dark background are very hard to read.

A lot of people have problems distinguishing red and green (and also, brown / green, blue / black, and blue / purple); so don't use those colors in combination. Avoid red as it has associations with negativity - it is the color often used by teachers to make corrections and in finance it indicates a loss.

The audience's ability to see your slides very much depends on the internal and external lighting of the room. If the sun is shining directly onto the screen it makes light colors (particularly yellow) almost impossible to see. Some beamers make red look like blue. Also, bright light considerably reduces the strength of color in photos.

8.9 Make your graphs come alive

The statistics that you give the audience will be probably be familiar to you, so you will have a natural tendency to explain them too quickly and in too much detail. The secret is just to select a few and explain them in a way that the audience can understand.

If the statistics are in the form of a graph, it helps the audience to understand better if you explain what the quantities are on each axis and why you chose them. This gives the data a context and also allows you to add some personal details about how and why you selected them. Obviously, however, if the axes are self-explanatory there is no need to comment on them.

Think about how you might explain and comment on the graph below (Fig. 8.1).

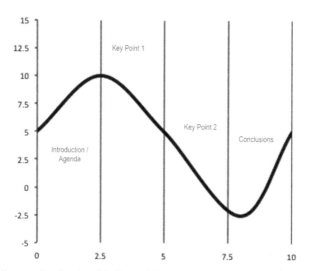

Fig. 8.1 Audience attention level (-5 to $+15$) over a ten-minute period

Would this be a helpful commentary for the audience?

In the graph that can be seen in this slide, which delineates the typical attention curve of an audience during a ten-minute presentation. As explained in the caption to the figure, the x axis of this two-dimensional plot of a sinusoidal curve represents the number of minutes, and the y axis the amount of attention paid by an audience. The graph highlights that at the beginning of a presentation the level of attention is relatively low. Then it rises rapidly, reaching a peak at about two minutes. After approximately three minutes, it begins to drop quite rapidly until it reaches its lowest point at around seven minutes thirty seconds. Finally, it rises quite steeply in the ninth minute and reaches a second peak in the last minute.

8.9 Make your graphs come alive (cont.)

The problem with the above is that it contains no information that the audience could not have worked out for themselves. Basically all you have done is describe the curve in a rather abstract and tedious way. What you really need to do is to interpret the curve and point out to the audience what lessons can be learned from it. You could say something like this:

OK, so let's look at the typical attention curve of an audience during a ten-minute presentation. [Pauses 3-5 seconds to let audience absorb the information on the graph]. What I'd like you to note is that attention at the beginning is actually quite low. People are sitting down, sending messages on their iPhone and so on. This means that you may not want to give your key information in the first thirty seconds simply because the audience may not even hear it. But very quickly afterwards, the audience reach maximum attention. So this is the moment to tell them your most important points. Then, unless you have really captivated them, their attention goes down until a minute from the end when it shoots up again. At least it should shoot up. But only if you signal to the audience that you are coming to an end. So you must signal the ending, otherwise you may miss this opportunity for high-level attention. Given that their attention is going to be relatively high, you need to make sure your conclusions contain the information that you want your audience to remember. So stressing your important points when the audience's attention will naturally be high - basically at the beginning and end - is crucial. But just as important is to do everything you can to raise the level of attention when you are describing your methodology and results. The best ways to do this are ...

Note how the presenter:

- does not describe the curve, but talks about the implications
- does not mention what the x and y axes represent because they are obvious in this case
- highlights for the audience what they need to know
- repeats his / her key points at least twice (i.e. give important information at the beginning and end, signal that you are coming to an end)
- addresses the audience directly by using *you*

9 PRONUNCIATION AND INTONATION

9.1 Create lists of your key words and learn their pronunciation

Using the correct pronunciation is more critical in a presentation / demo than in any other situation where you will use English. If you cannot pronounce the key words of your presentation correctly, the audience may not understand you and thus they will not be able to follow your presentation. Also, they may be reluctant or embarrassed to ask questions for clarification.

Pronouncing words correctly is thus fundamental.

A typical 10 minute presentation includes between 300-450 different words (depending on the incidence of technical terms and how fast the presenter speaks). The number of different words in presentations of 15 or 20 minutes does not usually rise by more than 10-20 words compared to a shorter presentation, since most of the key words tend to be introduced in the first ten minutes.

Of these different words, the majority are words that you will certainly be already very familiar with: pronouns, prepositions, adverbs, conjunctions, articles, and common verbs. In a typical presentation, the average person may need to use between 10 and 20 words that might create difficulty in pronunciation. Learning the correct pronunciation for such a limited number of words is not difficult.

You can identify possible problems with your English if, as suggested Chapter 4, you write a script.

So it is essential that you create a list of key words that:

• are contained in your speech / slides

• that might be used in questions from the audience

You should then learn the correct pronunciation so that you can say the word correctly but also recognize it when someone else uses it.

A. Wallwork, *Presentations, Demos, and Training Sessions*,
Guides to Professional English, DOI 10.1007/978-1-4939-0644-4_9,
© Springer Science+Business Media New York 2014

9.2 Use online resources to find the correct pronunciation of individual words

There are two main ways to find out whether your particular pronunciation of individual words is correct.

The first is to ask a native speaker to listen to you practising your presentation and write down every word that you pronounce incorrectly, and then teach you the correct pronunciation. This may be expensive and time consuming, but is very useful.

The second is to write out your entire speech (see Chapter 4) and then upload it into Google Translate (where you can use the listening feature) or Odd Cast (http://oddcast.com/home/demos/tts/tts_example.php). Oddcast is good fun because you can also choose the sex and accent (e.g. British, US, Indian) of the speaker. The pronunciation and stress of the automatic reader of these two free applications is accurate in most cases. But not all cases.

If you have any doubts, consult with an online speaking dictionary, which will clearly indicate where the stress should be. You can try howjsay.com, which even gives alternative versions of some words e.g. *innovative* and *innovative*, both of which are now considered correct (note: *innovative*, which is typical of many non-native speakers, is completely wrong and is not given as an option).

Other dictionaries, for example give you alternative pronunciations depending on the type of English - UK or US. For example Cambridge's excellent online dictionary:

architecture *noun* US ◀)) /ˈɑː.kɪ.tek.tʃər/ US /ˈɑːr.kɪ.tek.tʃɚ/ [U]

source: http://dictionary.cambridge.org/dictionary/british/architecture?q=architecture

Other useful dictionaries:

http://www.howjsay.com (British English)

http://www.learnersdictionary.com (US English)

http://oaadonline.oxfordlearnersdictionaries.com (US English)

By using such online resources you can:

- note down where the stress falls on multi-syllable words (e.g. *control* not *control*)

- listen for vowel sounds, and learn for example that *bird* rhymes with *word* and so has a different sound from *beard*

9.2 Use online resources to find the correct pronunciation of individual words (cont.)

- understand which words you cannot pronounce. This means that you can find synonyms for non-key words and thus replace words that are difficult to pronounce with words that are easier. For example, you can replace

  o a multi-syllable word like *innovative* with a monosyllable word like *new*

  o a word with a difficult consonant sound like *usually* or **thesis**, with a word that does not contain that sound *often, paper*

  o a word with a difficult vowel sound like **worldwide** with a word that has an easier vowel sound like *globally*

- make a list of words that you find difficult to pronounce but which you cannot replace with other words, typically because they are key technical words

- understand which sentences are too long or would be difficult for you to say

9.3 Practise your pronunciation by following transcripts and imitating the speaker

An excellent way to learn the correct pronunciation of words is to use transcripts of oral presentations. Many news and education corporations (e.g. bbc.co.uk and ted.com) have podcasts on their websites, with subtitles in English and / or a transcript. You can thus hear someone speaking and read their exact words in the transcript. You could practise reading the transcript yourself with the volume off. This will motivate you more strongly to listen to the correct pronunciation when you turn the volume back on.

In addition you can learn how to pronounce some phrases that are typically used in a presentation by going to: www.bbc.co.uk/worldservice/ learningenglish/business/talkingbusiness/

9.4 Learn any irregular pronunciations

Unfortunately English has a very irregular pronunciation system as highlighted in the table below.

PRONUNCIATION INCONSISTENCY	EXAMPLES
different word, same pron	would = wood; where = wear = ware; hole = whole; scene = seen
same word, different pron	*Live / laiv /* as in a *live concert; live /* liv */* as in *I live in Italy*
different spelling, same pron	man*age*, fr*idge*, sandw*ich*; for*eign*, kitch*en*, mount*ain*
same spelling, different pron	Were / here, cut / put, chose / whose, one / phone, high / laugh
silent letters	We(d)n(e)sday, bus(i)ness, dou(b)t, comf(or)table

9.5 Be very careful of English technical words that also exist in your language

A lot of English words have been adopted into other languages (and vice versa), often with different meanings but also with different pronunciations. Here are some English technical words and acronyms that are also found in many other languages: *hardware; back up, log in; PC, CD, DVD.*

Note that:

- words that are made up of two words have the stress in English on the first syllable: *hardware, supermarket, mobile phone*

- words whose second part is a preposition have the stress on the first syllable: *back up, log in*

- letters in acronyms have equal stress: *P-C, C-D, D-V-D*

It is a good idea to say key words and English technical words more slowly. This will give you the time to focus on the pronunciation, and the audience more time to hear / understand the words.

Give equal stress and time to each letter in an acronym. Remember that an acronym such as IAE is very difficult to understand because it contains three vowels, and vowels (and consonants too) tend to be pronounced differently in different languages. If you use acronyms in your presentation it is best to have them written on your slides too.

9.6 Practice the pronunciation of key words that have no synonyms

Imagine you want to say the following sentence, but that you regularly mispronounce the two words underlined: *Then I'll take a brief look at the underlying* architecture *and the* methods *we used.* Also, imagine that you cannot find synonyms for those three words.

The solution is to break down the word and identify which part is causing you problems. Let's imagine you are having difficulty with the last part of *architecture*, and you are pronouncing *-ture* as *tour ray*. Think of other words ending in '-ture' that you know how to pronounce that end in those letters: *picture, nature, culture, feature.* If you know how to pronounce those words then you can also pronounce the *-ture* in *architecture*.

Obviously you also need to be able to pronounce the first part of the word too. In this case it is useful to listen to an online dictionary (9.2) that pronounces the word for you. Try to transcribe the sound in a way that is meaningful for you: *ar ki tek cher.* Alternatively, if you are familiar with phonetics, then you can use the phonetic transcription: ˈɑː.kɪ.tek.tʃər

Now let's look at *method*, where the problem is typically in the first syllable. In this case find out the correct pronunciation. Then create a chain of familiar words that will lead you to the correct pronunciation: *get > met > metal > method.* You can then practice the difficult word by reading it in association with the familiar words. Using the same system would enable you to remember the difference in pronunciation between the initial syllable in *method* and *methane: see > me > meet > methane.*

Note: the same combination of letters may have a different stress or pronunciation, e.g. *method, methodology, methodological; photograph, photographer, photographic.*

9.7 Be careful of -ed endings

When you add -ed to form the past forms of a verb, you do not add an extra syllable. For example the verbs *focused, followed, informed* are NOT pronounced *focus sed / follow wed / inform med.* The number of syllables of a verb in its infinitive form (*focus*) and in its past form (*focused,* pronounced *focust*) is the same. The only exceptions are verbs whose infinitive form ends in -d or -t, for example *added, painted,* which are pronounced *add did* and *paint tid.*

9.8 Consider using a multisyllable word rather than a monosyllable word

If you want to use a word that contains a vowel sound that you do not find easy to pronounce, then use a synonym with more syllables. For example, from an audience' point of view it is easier to understand the word *difficult* than *hard*. In fact, if you mispronounce *hard,* you may produce a sound that is more similar to *hired* or *hoard* or *heard*. This could initially prevent the audience from understanding what you have said. Also, *hard* is a very short word. The audience does not have much time to hear it.

On the other hand, *difficult* has three syllables. If you mispronounce one of the syllables, then at least the audience has a chance to hear the other two, and thus a greater chance of understanding the word.

Clearly, the context will also help the listener. But generally speaking multisyllable words are easier to understand than monosyllable words (but see 9.9).

9.9 Use synonyms for words on your slides that you cannot pronounce

You can have words on your slides that you are unable to pronounce. However, when you comment on your slides, you can use synonyms (i.e. words with the same meanings). For example, you may be listing the advantages and disadvantages of a particular procedure. On your slide you write:

Advantages: a, b, c Disadvantages: x, y, z

Imagine that when you practice your presentation you realize that you cannot pronounce *disadvantage* easily. In fact, *disadvantage* has four syllables, so four chances of getting the stress wrong. Also, the final *-age* is not pronounced like *age* as in 'at my age' but like *-idge* in *fridge*. So there are many chances of you getting the pronunciation wrong. The solution is to write 'advantages and disadvantages' on the slide, but when you speak about the slide, you can say the 'pros are' or the 'good things are' and the 'cons are' or 'the bad things are'. As you say, 'pros' and 'cons' you point to the items on your slides, so that everyone can understand that you when you say 'pros' and 'cons' you mean 'advantages and disadvantages' - just in case someone in the audience is not familiar with the term 'pros and cons'.

9.10 Use your normal speaking voice, but don't speak too fast

Much of the success of your presentation will be in how natural you sound to the audience. A friendly informal tone is the most popular style for audiences.

So practise talking as if you were talking to a friend. This is not a skill you have to learn. You already have this skill – what you have to do is to remove the barriers that are inhibiting you from talking in your normal way. Barriers include:

* you think you have to adopt a specific 'presentation voice'

* you think it is not professional to talk in a normal colloquial way

* you are nervous and go into automatic presentation mode

If you speak too fast, it makes it difficult for the audience to absorb what you are saying. And the impression may be that if you are presenting information very fast then it is not particularly important.

Do not worry. No one is going to fall asleep if you speak slowly. In fact, your speed is probably faster than you think it is. Speaking slowly also gives you the time to focus on your pronunciation.

Many people speak fast in presentations because they are nervous. But if you speak slowly you will find that it automatically makes you feel much calmer, and you will thus give a better impression to the audience.

Some non-native English speaking presenters speak fast because they want to show the audience that they are fluent in English. This is rather an unhelpful approach because it means that those people whose English is not as good as yours will have difficulty understanding you.

9.11   Vary your voice and speed

If the sound of your voice never changes or you have a very repetitive intonation (e.g. at the end of each phrase your voice goes up or is significantly reduced in volume), the audience will lose essential clues for understanding what you are saying. You need to vary your:

SPEED   how fast you say the words. Slow down to emphasize a particular or difficult point. Speed up when what you are saying is probably familiar to the audience or will be easy for them to grasp.

VOLUME   how loud or soft you say the words - never drop your voice at the end of a sentence

PITCH   how high or low a sound is

TONE   a combination of pitch and the feeling that your voice gives

You can vary these four factors to show the audience what is important about what you are saying. You can create variety in your voice by

• imagining that there is a curtain between you and the audience - you are totally dependent on your voice to communicate energy and feeling

• listening to people who have interesting voices and analyze what makes them interesting

• recording yourself and listening to your voice critically

9.12   Help the audience to tune in to your accent

When you start talking you will not have your audience's full attention. They may be thinking about what they were doing immediately they sat down to listen to you (e.g. an important phone call they made, a meeting, an email). They may be texting a colleague or chatting to the person sitting next to them. So they are not fully focused.

Also, they may never have heard you speak before. Your accent may sound very unfamiliar. You can help them tune into your voice. You can do this by introducing yourself and giving them a bit of useful but non-key information. This means that if they don't understand every word you say in this part of your presentation, it does not matter. But it will allow the audience to settle into your presentation.

9.13   Use your outline slide to introduce key terminology

There will be certain key words that you will want to use throughout your demo. It is essential that the audience understands these words, otherwise they risk not being able to follow the main thrust of your presentation.

An Outline slide is a useful way of introducing key terminology, as in the words in italics in the slide below.

OUTLINE

➢   Modification of *polymeric* materials

➢   *Bioreceptor-surface coupling*

➢   Characterization of *functionalized surfaces*

Give your audience a chance to tune into your key words immediately at the beginning of your presentation. This will help them to understand you in the rest of the presentation.

So you could put up the slide and say:

*So here's what I will be talking about. [Pause for two seconds so that audience can absorb the content of the slide] I first became interested in modifying **polymeric** materials because .... Then one day we decided to try **coupling** the **bioreceptors** with the activated **surfaces**. So those are the two things that I will be looking at today, along with some approaches to characterizing **functionalized surfaces.***

The benefits are the audience will:

• see and hear you say the key words and thus be able to connect your pronunciation with the words on your slide

• familiarize themselves with your voice without missing any vital information (you have simply told them why you are interested in this topic).

If you are still worried that people will not understand your pronunciation, you can point to the key words on the slide as you say them.

9.14   Use stress to highlight the key words

When you stress a word in a sentence you:

- say the word more slowly (and perhaps a little louder) than the ones before and the ones after

- raise the volume of your voice a little

- give your voice a slightly different tone or quality

Stressing particular words or phrases helps the audience to distinguish between non-essential information (words and phrases said more quickly and with no particular stress) from important information (said more slowly, with stress on key words).

Try saying the following sentences putting the stress on the words in italics:

Please *present* your report next week. (present rather than write)

Please present *your* report next week. (your report not mine)

Please present your *report* next week. (your report not your marketing ideas)

Please present your report *next* week. (not this week)

Please present your report next *week*. (not next month)

Although the sentence is exactly the same, you can change the meaning by stressing different words. And the stress helps the listener to understand what is important and what isn't.

9.15   Enunciate numbers very clearly

You can help your audience by writing any important numbers directly onto your slides.

Also, remember to use the correct pronunciation to distinguish clearly between *thirteen* and *thirty*, and *fourteen* and *forty* etc. Note where the stress is: *thir**teen** **thir**ty*. Make sure you enunciate clearly the *n* in *thirteen, fourteen* etc.

If you use the correct pronunciation, the audience will be able to see the difference in the shape of your mouth. To pronounce -*een* your mouth will be wider, to pronounce *thir*- and *four*- you will need to push your lips forward and form a round shape with your mouth.

9.16   Avoid *er, erm, ah*

In order not to distract the audience, try hard not to make any non-verbal noises between words and phrases. You may not be aware that you make these noises. To check if you do, record yourself delivering the presentation.

You can stop yourself from saying 'er' if you speak in short sentences and pause / breathe instead of saying 'er'.

In any case if you practice frequently you will know exactly what you want to say, so you will not have to pause to think. Consequently there will be no gaps between one word or phrase and another, and thus no need to say 'er'.

9.17 Mark up your script and then practice reading it aloud

When you have created a final version of your script (see Chapter 4), you can mark it up as shown below. You probably won't have time to do a full mark up for your whole presentation. But it is important that you do it for your introduction, which is the time when the audience is tuning in to your voice and making their first impressions. You should also do it for your conclusions. Also, it is a good idea to mark all those words that a) you intend to give EMPHASIS to b) those words that you find difficult to pronounce.

> *Thank you **very** much for **coming** here today. / My name's **Esther Kritz** / and I am the technical **manager** here at **Soft Mick** inc. Previously [pre = he]/, I spent **three** years at **Microsoft** / and before that / I was one of the **first** programmers at Google. / / I'd like to show you / what I think / are some INCREDIBLY useful features of ...*

SYMBOL	MEANING
slash (/)	Indicates where you want to pause. You only need to do this for the first 30-60 seconds of your presentation. A typical problem of the first seconds of your presentation is that you are nervous and this makes you speak very fast. If you speak too fast the audience has difficultly understanding. If you insert pauses this should encourage you to slow down and also to breathe. By breathing more you will become more relaxed.
double slash (//)	A longer pause. If you pause between key phrases it will focus the audience's attention on what you are saying and also give them time to digest it. Long pauses can have a positive dramatic effect.
bold	Words that you want to stress in each phrase. This does not mean giving them a lot of stress but just a little more than the words immediately before and after. This stops you from speaking with a monotone (i.e. with equal stress on each word) which is boring for the audience. Words that tend to be stressed are key nouns, numbers, adjectives, some adverbs (e.g. *significantly, unexpectedly*) and verbs. Words that are not generally stressed are pronouns (unless to distinguish between two entities - *I gave it to **her** not to you*), non key nouns, prepositions, conjunctions and most adverbs.
CAPITALS	Words that you want to give particular emphasis to. You may want to say them louder or more slowly or with a particular tone of voice. You do this to draw the audience's attention to what you are saying. Words that tend to be given extra emphasis are numbers and adjectives.
underline	Highlights the primary stress within a particular word.
[ ]	Insert in brackets the sounds of words or syllables. For example, if you write *pre = he* this will remind you that the sound of *pre* is the same as in *he* rather than the sound as in *present*.

9.18   Reading your script aloud during the presentation

If you are very nervous about giving a demo or presentation in English, then one solution is to read it aloud.

I do not mean reading it aloud from a piece of paper. Instead:

1. write a script of everything you want to say (see Chapter 4) and have it checked by a native English speaker

2. mark the script as suggested in 9.17 (insert it into Google Translate or similar to check the pronunciation - see 9.2)

3. divide up the script so that each slide has its own script

4. insert the scripts slide by slide into the presenter's notes in your presentation software

5. transfer your presentation onto a tablet or mobile phone.

During the presentation, you can then hold the tablet in your hand and read directly from the notes. The trick is to try to maintain eye contact with audience as much as possible. This means that you should not focus only your eyes on the tablet. Instead, you should try to learn as much as possible what you want to say and thus make it seem that you are not reading.

There are huge advantages to this solution:

• you will feel much less nervous and much more in control

• you can mark the pronunciation of difficult words in your script

• you will probably speak more slowly

• you will not improvise, so you will make far fewer mistakes and it will also help you to respect the time available

• the audience will probably understand you much more clearly

You will be surprised at how much extra confidence you will gain by knowing that you can glance at your tablet if you need to. In reality, many people who use this technique never actually need to look at their tablet!

This solution is obviously not the optimal solution. Much better is to conduct your demo or presentation without reading it directly from a tablet. However, this is a great solution if you are anxious about speaking in public or have had no time to practise.

10 USAGE OF TENSES AND VERB FORMS

10.1 It's OK to make mistakes in the main body of the presentation

Don't let English become the main focus and problem of your presentation. Always put content before English: if your content makes your message clear, a few mistakes in English will make no difference.

Misuse of grammar in a technical presentation rarely causes problems. When you are explaining technical things, the audience will be more focused on what you are saying rather than how you are saying it. The audience is made up of clients or colleagues most of whom will be motivated to see your demo – they are not English teachers wanting to assess your linguistic proficiency. The way you relate to the audience and involve them, is more important than any grammatical or non-technical vocabulary mistakes that you may make.

You only need to use a limited number of constructions and tenses.

If you make an English mistake while doing your presentation:

- don't worry (the audience may not even notice)

- don't correct yourself – this draws attention to the mistake and will distract you from what you want to say from a technical point of view.

If you are giving presentations in English where the majority of the audience speaks your native language:

- choose words and expressions that you know your audience will be familiar with

- don't try to show how well you speak English by speaking very fast – your aim is to help them understand

A. Wallwork, *Presentations, Demos, and Training Sessions,*
Guides to Professional English, DOI 10.1007/978-1-4939-0644-4_10,
© Springer Science+Business Media New York 2014

10.1 It's OK to make mistakes in the main body of the presentation (cont.)

Tenses and verb forms are used in different ways in different parts of the presentation. The most frequently used are:

present simple: *I look*

present continuous: *I am looking*

present perfect: *I have looked*

present perfect continuous: *I have been looking*

past simple: *I looked*

future simple: *I will look*

future continuous: *I will be looking*

going to: *I am going to look*

conditional: *I would [like to] look*

imperative: *look, let's look*

infinitive: *to look*

gerund: *looking*

You can always either use full forms (e.g. *I will, I am*) or contracted forms (e.g. *I'll, I'm*). There is no difference in meaning, but the full forms can be used for emphasis, and the contracted forms sound more informal.

You don't need to have a perfect understanding of English grammar in order to be able to use the tenses correctly. The examples of tense usage in this chapter are in the form of useful phrases that you can say at particular moments during your presentation.

10.2 Getting to know the audience (small demo)

Present Simple: habitual situations.

*So what exactly **do you do** at ABC?*

*I **work** in sales.*

Present Perfect Continuous: from past to present

*So how long **have you been working** for them?*

*I **have been working** in this group since 2013.*

Present Continuous: current / temporary activity

*So what **are you working** on at the moment?*

*What **are you hoping** to learn from this training session?*

10.3 Webcast introduction / Large-scale very formal presentation

Present Simple: scene setting / introduction.

*It **is** now 3 o'clock and it **gives** me great pleasure to welcome you to the second annual general meeting of ABC. **I am** delighted to see so many shareholders here today. We **appreciate** your interest in the company and the support that you **continue** to give us.*

*The presentation **is** about 45 minutes and then we **have** about another 30–40 minutes for Q&A after that.*

Imperative: giving instructions.

*One small reminder – pleases **witch off** your mobile phones.*

*If you would like to be placed on the mailing list, please **let** us know.*

Would like: introducing polite wishes.

***I'd like** to welcome you to this technical briefing on the ...*

*Before we start, **I'd like** to draw your attention to two items.*

10.4   Outline / Agenda – informal demo or presentation

Three tenses are usually used in outlines. When you outline your first point, just use either *going to* or the future continuous. For the other points, you can also use the future simple.

As you can see in the examples below, you can use a variety of future tenses. However, do not use the present continuous, and don't use **will** for the first thing you say. If you use **will** for the first item, then it gives the presentation a very formal feel. Also, **will** often gives the idea of a decision taken at the moment of speaking, so it may seem that you are improvising.

*Let me just outline what **I'll be discussing** today.*

*First, **I'm going to tell** you something about the differences between the old and new versions.*

*Then **I'll take** a brief look at the most important enhancements.*

*Finally, **we'll be doing** a bit of hands-on practise.*

If you want to leave some options open to the audience you can use *am planning*.

*What I **am planning** to do today, if it is OK with you, is to ...*

*I **am not planning** to deal with X, because ...*

Here are some more examples with other tenses:

*Today's presentation **covers** five topics. These **are:** First, the adoption of a new standard for ... Second, the release of:*

*Today I **will cover:** Why we **think** there **are** interesting opportunities beyond our current footprint. Why we **think** we **are** the most attractive partner as **we** look to source transactions and partnerships. What outcomes we **have set** ourselves for growth.*

10.5 Outline / Agenda – formal presentation or webcast (possibly with several speakers)

will / shall: explaining how the presentation has been organized.

> *In a moment I **shall say** a few words on the company's results. Ruiki Yamashata, our Chief Executive, **will then update** you on our performance so far this year and talk about progress on our strategic priorities. After that, we **will move** to the formal business, and I **will invite** you to put forward any questions you may have. We **will then vote** on the resolutions. I **shall** then ...*

> *After our prepared remarks, Sara and I **will be** pleased to take your questions on the technical changes shown here.*

> *You **will find** all of the information from the slides, along with further supporting detail, on our website.*

Future Continuous: explaining how the presentation will proceed.

> *We **will be referring** to slides that we **will be using** during the webcast. If you are listening on the phone, these are available to download from our website.*

would like: outlining what you will do next.

> ***I'd now like** to introduce Vladimir Raspovic, our Chief Financial Officer.*

> *First, **I'd like** to take a step back and review ...*

10.6 Giving background details

Past Simple: events and situations that have finished.

> *We **started** working on this in **May last year.***

Present Perfect: open issues, progress so far – the precise time is not important.

> *We **have already developed** two prototypes.*

> ***Since** the market first opened, we **have made** a number of modifications.*

> *We **have added** several new features in this version.*

10.7 Presenting financial highlights

Present Perfect: announcements, general trends where no specific time is mentioned.

*Revenues **have increased** for the third consecutive year.*

*We **have opened** up two new offices in the Middle East.*

*We **have gained** a bigger market share than our expectations.*

Past Simple: finished events (e.g. looking at the past financial year).

*The business **achieved** a further year of operating profit growth, up 2 % to € 2.4 billion. Earnings **increased** by 3 %, to £ 1.3 billion, equivalent to 25.8 cents per share.*

*This **resulted** in an effective group tax rate of 40 %.*

10.8 Describing your company

When describing your company you will primarily use a mix of the Present Simple and Past Simple, both in the active and passive forms.

*Thanks for the opportunity to tell you something about BejPharm. BejPharm **is based** in Beijing and **was incorporated** in 2009. We **filed** to go public in 2013, and we **started** trading on the Shanghai Stock Exchange in 2014. Our market capitalization **is** about 300 million yuans. We **have** a very clean capital structure, 51 million shares outstanding, and 21 million under float.*

*We **have** two different divisions, a pharmaceutical division that **focuses** on neurology, and a second division, our genetics division, which **started** three months ago and is already generating revenue.*

*All of our technology **is patent protected** and **is based** on our knowledge on how to optimize gene expression.*

*We **have** an experienced management team. We also **believe** that we **have** a very different business structure, which has allowed us to ...*

*At the moment we **have** a workforce of 43,000, most of whom **are based** in China. Last year we **opened** up offices in Central Africa, and we **plan** to ...*

In the above example notice also the use of the:

- Present Continuous to indicate an action or state that is taking place now (second paragraph: *is already generating revenue*).

- Present Perfect to indicate the consequences of a certain action or state (penultimate paragraph: *which has allowed us to...*)

10.9 Describing your role in the company

Past Simple: to describe previous activities that have now finished.

*Just to give you a bit of my own background. I **did** my Ph.D. in International Finance at the University of Wuhan and **began** my career at the Industrial and Commercial Bank of China. Then I **moved** on to Accenture, where I **headed** up the financial products department. I **joined** ABC in 2012, at that time they **needed** someone with more product experience. I **became** the CTO in 2014.*

Present Simple and Present Continuous to describe what you are doing now, Present Perfect (Continuous) to describe activities that began in the past and are still active now.

*My name is Yudan Whulanza and I **am** a senior developer here at ABC. **I've been** in the role for four years, responsible both for strategy and business development. In total I **have been working** with the company for ten years. At present we **are developing** a series of new services to offer our clients.*

10.10 Referring to future points in the presentation

Use either the future simple or the future continuous. In this context, there is really no difference in meaning.

*As we **will see** in the next slide ...*	*As we **will be seeing** in the next slide ...*
*I'll **tell** you more about this later ...*	*I'll **be telling** you more about this later ...*
*I **will give** you details on that at the end ...*	*I **will be giving** you details on that at the end ...*

Don't use the Present Continuous to refer to future parts of the presentation. Only use it when informing the audience about what you are doing now or when hypothesizing about what they are probably thinking as they see the slide.

*I **am showing** you this chart because ...*

*Why **am I telling** you this? Well ...*

*You **are** probably **wondering** why we did this, well ...*

10.11 Answering audience questions during the presentation

If you reply to a question during your presentation (i.e. <u>not</u> in the Q&A session at the end), for a native speaker there is a difference in meaning between *will* and the Future Continuous. *will* may give the idea that you have made up your answer on the fly, whereas the future continuous gives the idea that you had already anticipated that particular question (even if you hadn't!). For example:

Audience question: Can you tell us the website address?

You: I will be giving the address at the end.= Don't worry I had already thought about this and have prepared a specific slide with the website address on it.

You: I will give you the address at the end. = I am not a very good presenter/organizer, I hadn't thought about telling you the address but I have decided now to tell you it at the end of the presentation.

In the two cases below *will* is correct because you are responding spontaneously to a request from the audience. In these cases, your organizational capacities are not going to be doubted.

*Let me think about that and I **will tell** you at the end.*

*OK **I'll just go back** to the last slide to explain that again.*

10.12 Mentioning the design and development phases of a product or service

The design and development stages are generally finished when you are presenting a product so use the Past Simple.

*I **used** Java.*

*We **sent** clients a questionnaire.*

*We **concluded** that the best way to do this was ...*

10.13 Talking about how your prepared your slides

When you refer to the choices you made when preparing your slides, use the Present Perfect.

*I **have included** this chart because ...*

*I **have removed** some of the results for the sake of clarity ...*

*I **have reduced** all the numbers to whole numbers ...*

10.14   Making transitions

When you refer to what you have done up to this point in the presentation, use the Present Perfect. This is often used for making mini summaries before moving on to a new point.

*So we **have seen** how X affects Y, now let's see how it affects Z.*

*I **have shown** you how this is done with Z, now I am going to show how it is done with Y.*

*So, let's just look at where we are in the agenda. OK. **I've covered** the opportunities and why we think that … So now let's talk about outcomes.*

But when you are talking about moments earlier in the presentation use the Simple Past or *may*.

*As we **saw** in the first / last slide …*

*As I **mentioned** before / earlier / at the beginning …*

*As **noted** earlier …*

*You **may** recall from an earlier slide that …*

To move to the next slides you can use would *like, let's,* various future forms and the Simple Present.

***I'd now like** to turn to our second major innovation.*

***I'd like** to now move from cost-related matters, and cover several other changes.*

*The next few slides **cover** the four reasons why we think …*

*Now **I'm going to introduce** a new topic: …*

*Next **we'll be looking** at how your company can benefit from these changes.*

*OK. **Let's now look** at some detailed figures.*

10.15   Highlighting what is on a slide

Don't be afraid to use the imperative to instruct your audience what to do – it is not impolite. Alternatively you can use a conditional form.

***Note** how this line increases sharply here.*

***Remember** what I was saying earlier about XYZ …*

***I'd like** you to focus on this part of the slide.*

*If you **take** a look at this chart **you can / will see** that …*

10.16   Explaining figures, tables, charts and diagrams

You can use the Simple Present (active or passive) to indicate something that is certain.

*This slide **shows** ...*

*The dotted lines **represent** ...*

*Time **is represented** on the X axis.*

*This **is** a detail from the previous slide.*

Use *should* and *supposed to* when it is not so certain and could be open to interpretation.

*This table **should** clarify ...*

*This picture **is supposed to** represent ...*

10.17   Indicating level of certainty

Some presenters like to sound tentative when talking about performance. This makes them sound less arrogant and also means that they leave the way open to other interpretations.

*Our tests **would seem to** indicate that*

*This increase in performance **should help** you to ...*

Various tenses and modals indicate different levels of certainty (the percentages are only indicative):

Present Simple: 100 % certainty (The product **works** well.)

must (not), cannot: 100 % (The product **must not** be used outside laboratory conditions.)

should: 90 % (The software **should** be released next month.)

might / may / could: 50 % (This **may** cause damage.)

10.18   Conclusions: formal presentation / webcast

You will need to use a variety of tenses and forms during your conclusions.

*In summary, the changes **we are making** next year **should lead** to ...*

***To conclude, let me say** that ...*

*In conclusion, the forecast for the next decade **is looking** very bright.*

***I'd now like** to pass the presentation to Lee for some closing remarks.*

*We **hope** this is useful to you in your work. My colleagues **would be happy** to take you through any of this material off-line.*

*That **concludes** our prepared remarks. **We'd now be pleased** to take your questions.*

10.19   Conclusions: demos and less formal presentations

You will need to use a variety of tenses during your conclusions.

*OK. So basically you **will have** three main advantages in using this new version:*

*First you **will be able** to do xxx ...*

*Second you **will save** time because you can ...*

*Finally, as we **have seen**, the new version is considerably faster and more flexible ... This **means** that ...*

10.20   Q & A Session

To learn how to deal with questions from the audience see Chapter 13 and Chapter 15.

could: a polite requests

> *Sorry,* **could you repeat** *the question more slowly please?*

> **Could you speak** *up?*

Simple Past: to refer to the act of hearing or understanding the question.

> *Sorry, I* **didn't hear** *the last part of your question.*

Present Simple: to indicate continued lack of understanding.

> *Sorry,* **I still don't understand** *– would you mind asking me the question again in the break?*

Present Perfect: to indicate connection with the present situation.

> *You* **have raised** *an interesting point, which, to be honest, I have never thought about before.*

would: to delay answering the question.

> *Sorry, but to explain that question* **would take** *rather too long, however you can find the answer on our website.*

> **Would you mind** *sending me an email with your question?*

> **Would you like** *to have a coffee after the presentation and we can talk about this in more detail?*

11 CONDUCTING A PRESENTATION, DEMO, OR TRAINING COURSE WITH A FACE-TO-FACE AUDIENCE OR VIA VIDEO CONFERENCE

11.1 If some participants arrive early, exploit the opportunity to ask them questions.

It helps if you know as much about your audience before you begin, then you can tailor it to their needs.

You can ask them some fairly generic questions about their role in their company and how long they have been there. Listening to their answers will give you an opportunity to get used to their voices a little. Then during the demo when they ask you questions you should be in a better position to understand them. Typical questions (social, work, technical) you can ask participants include:

Have you just flown in or did you spend the night here?

Do you live round here?

Can you tell me what your role is inside the company?

Do you have any knowledge of the topic of today's demo already?

A. Wallwork, *Presentations, Demos, and Training Sessions,*
Guides to Professional English, DOI 10.1007/978-1-4939-0644-4_11,
© Springer Science+Business Media New York 2014

11.2 Get to know your audience when they are all present

When everyone has arrived, find out who they are (see 3.3 and 3.4 to learn how to find out about participants in advance), what they might be expecting from you, and what they are interested in.

The idea is not to give them a standard message, but a message that seems tailored for them. Get agreement on their expectations and priorities.

Then, tell them what you are planning to do:

1. when you have an agenda that you really need to follow:

   Before we actually start I just wanted to run through what I had planned for today. So I'm going to start with an overview of X. Then I'll move on to Y, and then ... The whole thing should take about 30 minutes. If you want to take notes, that's great, but I will be giving you a handout after the presentation. Also it would be good if you could ask questions at the end. How does that sound?

2. when you are open to them driving the presentation:

   Before we actually start I just wanted to run through what I had planned for today. So I was thinking of starting with an overview of X, because I think most of you are not that familiar with it. Then I'll move on to Y, and then I thought perhaps we could have a short training session. After that we can have a break for lunch, and then continue the training ... The whole thing should take about 4 hours. Does that sound OK, or are the some things you want to leave out or do in a different order?

11.3 Before you start find out the names of the people in your audience

If you are presenting to a small group of people make sure you find out and remember their names. You can do this by getting each one to introduce themself before you begin. You repeat their name immediately. Example:

Participant: Hi, I am John Smith, and I work in the IT department.

You: John, glad you could be here. Can you tell us a bit more about yourself.

Then you note down their name on a piece of paper. Alongside the name put some physical characteristic that will help you to identify them (e.g. glasses, long hair, beard, bald).

If you think you might have difficulty understanding their names, then either:

* before the presentation get someone to email you the list of participants (see 3.3 for other reasons why having an attendance list is useful)
* get them to write down their names in capital letters on a piece of paper – depending on the level of formality of the presentation / demo, you could just get them to write down their first name

During your presentation you can then address people directly by name:

John, is this a problem that you have encountered?

or more indirectly by profession:

For IT people like yourselves ...

As you know this is something people in IT frequently encounter ...

From what I have gathered in previous sessions, most IT people find that ...

Basically, people like to hear their own name, they feel special and they are naturally move motivated to listen to you.

11.4 Position yourself and your laptop where your audience can see you

Decide the best position to locate your laptop in relation to the audience and the screen (see 16.2). You need to be able to see the audience clearly and they need to see you as well as the screen.

Stand up as much as possible (see 16.3). When you can, operate your laptop remotely.

11.5 Give a clear signal that you are going to begin, then give your agenda

Although you've negotiated with your audience the content and structure of the presentation, you still need to make a clear signal that you are now going to begin the presentation.

> *Right, I think we better get started. First, let me just give you the agenda for the first part of the presentation, and remember if you have any questions please feel free to interrupt me. In any case I'll be giving you a handout at the end. So, ...'*

Don't begin the demo still sitting down behind your laptop. Remember you are representing your company and you want to create a good image. So:

- blank the screen (i.e. so that nothing is being shown)
- stand up
- thank everyone for coming
- introduce yourself
- say why you were chosen to give the presentation / demo
- outline the agenda
- underline why what you are going to say will be interesting for them
- say how long the presentation will be and when there will be breaks
- tell them when they can ask questions
- inform them whether there is a handout and whether they can refer to it during the presentation.

Here is an example of how to begin a training session to colleagues from your own company:

> *Hi, I'm Miroslav Zawinul and I am responsible for XXX training. My main role is customer support. My strength is on the business side. As you know, you are already using one of our products – JAVA XXX. I'm here today to introduce you to the new, totally revamped version – version 120.*
>
> *What I am planning to do, if it's OK with you, is first to go through the technical enhancements in the engine because ... Then I am going to introduce you to the new XXX, highlighting the differences with the old one etc*

11.5 Give a clear signal that you are going to begin, then give your agenda (cont.)

I think it's really important for you to be here. You need to be involved, as it is essential for you to understand the rationale behind the product so that you can convince, or at least support, customers in moving from the old system to the new one.

I am assuming that you already have quite a bit of knowledge of Java. But do feel free to stop me and ask any questions. At any point we can go back and look at the protocols either briefly or I can do a separate presentation on them.

If you don't have any questions, the first part should only take five or ten minutes. Then the second part should be just over an hour. If I can't answer a particular question I will certainly get back to you by email with an answer.

So that's the agenda. Oh, by the way, do you all have the handout? It should help you follow what I am saying, and you will see that it also contains a lot of further notes that you can look at later.

OK. So now I'm going to ...

11.6 Tell them how they will benefit from the demo and also what you are not going to do

Explain what you are going to do and tell them how it relates to their work and how it will benefit them. They will be more motivated to listen to you if you tell them what they will be able to do at the end of the demo / training session that they can't do now.

You may also like to tell them what you are <u>not</u> going to talk about.

There are going to be four blocks of training. Today we're going to do ... Then on Wednesday and Thursday we'll talk about... I won't be discussing x because... And unless you particularly want me to, I haven't planned to cover ...

11.7 Optimize your agenda when your demo shows an update or new release of a product / service

Frequently demos are designed to update colleagues and clients on new features of a product or service. Clearly, before you begin your demo you need to find out whether the participants have used and are familiar with the previous version. You can have an informal chat about what they liked and did not like about the previous version – make sure you don't just focus on the negative. Hopefully, the elements that they did not like in the previous version will have been fixed in the new version – and you can certainly tell them this immediately. This will then make your audience more stimulated to listen to your demo.

Make sure you note down the elements / components / features that the audience did not like in the previous version, along with the name of the person who expressed dissatisfaction. Then when you say how these features have been fixed, you can look at the person who expressed dissatisfaction and ask what they think about the updated version. This is obviously risky, if they are not happy with the new version either, but even in this case you can say:

Well, what you have said is very interesting, I will refer it to the development team and see if they can implement what you are suggesting.

This shows to your client or colleague that you (and your company) are genuinely interested in getting feedback.

After this initial chat, you can then show your agenda. The agenda could include more items that you actually plan to cover and you can say:

Don't worry, this is just a full list of the features we have changed, we are not going to cover all of them. What I would like to know is which features you would like me to walk you through.

This is very empowering for the audience as they will feel that they are deciding themselves what your demo will be about. If they choose the features to cover they will naturally be more interested in hearing what you have to say about them. This is vital for you as maintaining your audience's level of attention is very challenging, particular when the demo / training sessions lasts for several hours or even days.

On the basis of the features that the audience choose, you can then begin your demo.

11.8 Begin your demo in a dynamic way

With your very first words you need to attract your audience's attention. Compare the two versions below:

ALIENATING AND REDUNDANT	ENGAGING
Good morning and thank you for finding the time in your busy schedule to attend today's demo. First of all I'd like to say a couple of words about myself, my name is Heinz Winkel, I am one of the account managers here at Schmikel Industries. I'm here to talk to you today because I would like you to see the new version of our product Turbo Schmick. This new version has the capacity to lead to an increase in the rapidity with which working processes are executed of around 300%.	*I am very pleased you could come today. My name is Heinz Winkel, I am one of the account managers here at Schmikel Industries. I'm here to give you a quick overview of an exciting new version Turbo Schmick. Turbo Schmick 3.0 will speed up your working process by about 300%.*

If you begin your demo as in the first column, the message that you are giving to your audience is that they will be forced to listen to a lot of words with little content. It is an invitation for them to go to sleep. Remember that we are conditioned by initial impressions. If our first impression is that we are going to get very little value from listening, then we are going to be less motivated to follow the demo. If the presenter then continues as in the first column below, then he/she will have already lost the audience's attention.

ALIENATING AND REDUNDANT	ENGAGING
What I am going to do is to give you a quick overview about the structure of this product and the main features that have been introduced in this new release with a comparison of version 2.0 that you are using now in order to highlight the main pros and advantages offered by this new product.	*There are three main features that I think you are really going to be interested in. First .... Second ... And finally ...*

11.9 Motivate your audience: don't focus on what they already know and don't sound negative

You should start off with a little of what the audience already know, so as to make them feel comfortable. But while you are doing this, avoid phrases such as:

What I am going to tell is quite interesting.

You may have already heard these explanations before.

You've probably already done this kind of training a million times.

I imagine you've already seen this picture / diagram / figure before.

The above phrases will simply demotivate your audience. They are a signal that what you are going to tell them is nothing new. The word *quite* in the first phrase could be interpreted as 'not particularly interesting' – you need to sound as enthusiastic as possible.

Basically never say anything that could be interpreted as being remotely negative.

Focus on:

- sounding positive
- being as charismatic and informal as possible
- using simple and concrete explanations
- not reading your slides karaoke style

As quickly as possible you should move on to new territory. Tell the audience:

- how what you are going to teach them will change their working lives
- new features that make your application different from previous versions or from rival products

11.10 Show a slide. Then pause before you begin talking

When you show a slide, pause two seconds to let the audience start reading. Give them time to assimilate the information and then draw the audience's attention to the importance of the information being shown in that slide.

Most audiences find it difficult to read and listen at the same time.

11.11 Use *you* not *I*

Try not to use *I* or *we* too often. Where possible express things in terms of the audience by addressing them directly. So instead of saying:

> First, **I will teach you** how to do x. By the end of the session I hope to **have shown how to do** x and y.

You can say:

> First, **you will learn how** to do x. ... Then by the end of the session **you should be able to** do and y.

This technique should not be used all the time, but will certainly make the audience feel much more involved and more motivated to listen to what you are going to tell them.

11.12 Explain components and features in terms of how the audience will use them, but don't explain the obvious

When you describe the features of your product or service, it helps if your explanation can answer the following questions that your audience might ask (or will at least will be thinking):

- Very briefly, what is this feature?

- Why is it useful? What problem does it solve? What will it enable me to do that I couldn't do before?

- How do I use it and in what context?

- How will it speed up my work or those of my colleagues / clients?

Avoid obvious explanations. Certain features are not worth explaining, for example the use of a *reset* button. So avoid saying:

> There is also the reset option <u>which enables you to reset the xyz.</u>

Such explanations are:

- redundant – clearly any technical person knows what *reset* means and does

- annoying and boring – it will simply make the audience lose their concentration and they may miss the important things that you have to say later on

If you must mention something so obvious, then say it very simply with no explanation.

e.g. *You've also got 'reset'. Then, on this side there is ..*

11.13 Move your cursor slowly around the screen

One of the worst mistakes you can make is to move your cursor around the screen too fast for the audience to follow what you are doing. Remember that you know what you are doing and you have practiced doing it many times. However, for the audience this will be the first time. If you go too fast the audience will become frustrated and may simply lose all interest.

11.14 Focus on examples, not on theory

Imagine you are attending a training course on the way financial markets work and all the tricks traders use to buy and sell commodities. The trainer tells you that she is now going to talk about 'Call Options' and shows you this slide.

What is a call option?

"A CALL option is the RIGHT but not the obligation to BUY a fixed quality and quantity of an underlying asset at a fixed price at or before a specific date in the future."

You have already seen five or six other slides with similar definitions for other types of Options. So how motivated are you going to be to try and understand what the trainer is telling you? Not very.

But what if the trainer instead began not with the definition (i.e. the theory) but with an example. In this case instead of showing a slide with a definition, she just shows you a picture of a house, with a photo of man on the right, and a woman on the left. You are naturally going to want to know why there is a picture of a house on the screen, so you will be motivated to listen. The trainer says:

Suppose Bob [trainer points at the man] *has a house that he wants to sell and which is worth 100,000 euros.*

Mary [trainer points at the woman] *thinks the value of the house is likely to go up. She wants the option to buy the house in one year from now for 100,000 euros. But she doesn't want to be forced to buy it, she only wants the right to buy it. So Mary pays Bob 10,000 euros to sign a contract to give her this right.*

This kind of contract is called a call option.

Two very important things to note from this example:

1. The example is about houses. Everybody can relate to house values, they are something that affect the majority of people.

2. The sentences are short and concise. The trainer can pause naturally between one sentence and another. This helps the audience to absorb the information step by step.

The trainer can now give the formal definition of a call option. By showing the definition after, rather than before, the example, the audience has a much greater chance of understanding and retaining the information that they are being given.

So, where possible given an example before explaining the theory.

11.15 Don't overload the audience with too many concepts

So what should the trainer do next: move on to the next type of Option, or give more details about the Call Option?

A key rule of training is that an audience can only understand and remember a finite amount of information – the less information you give them, the better they will remember it.

When training, you cannot teach everything you know. This will overwhelm the audience and demotivate them. Instead concentrate on a few key points. They will then be stimulated to learn more about the topic themselves – and you can tell them where to look for this information, or even provide it yourself.

So to return to our question, the trainer should probably give more information about the Call Option – but not more theoretical details. You need to tell your audience why what you have just taught them is important, and if possible how it might impact on their own work or life. So in this case, the trainer could tell the audience how and why investors exploit Call Options (i.e. for speculation and hedging).

Moral of the story: Always tell your audience why something matters.

11.16 Learn how to gauge the audience's reactions

Try to get feedback from their faces and body positions on how much they are understanding and how much they are interested. If the audience is becoming impatient, perhaps you are spending too much time on something which is fairly obvious for them. Skip slides or go faster / slower accordingly. Ask yourself:

- Are the audience alert? Are they following me? (they are maintaining eye contact with you)

- Are they interested? (they are nodding their heads in agreement or understanding)

- Is someone wanting to ask a question? (sitting forward, mouthing words)

128

11.17 Constantly elicit audience feedback on how to improve your product and services

One of the main aims of both internal presentations (e.g. with sales and business development) and external presentations (e.g. with customers) is to get suggestions, for example, for new features of an existing product / service or a totally new one that is still in the development stage.

To get such feedback, you need slides that clearly show the new features etc, but most of all you need to ask questions. You cannot simply ask *What do you think of x?* or *Would x be useful*? as the audience could just say *yes* or *no*. You thus need to ask them more open questions such as

Alright, so I know you like this feature, but how could it be optimized?

During your preparation identify which of your slides you could use as a springboard for your own questions.

11.18 Responding to attendees' feedback

You must always acknowledge feedback from the audience in a positive way – even if it has not been particularly useful. Never just move straight to another person.

TYPICAL IDEAS FROM THE AUDIENCE	EXAMPLE RESPONSES
New ideas that you have never thought of before	*Brilliant! That sounds like something really useful, thank you.*
Ideas you already have and that are currently being investigated	*Yes, that's great, and actually it's something we are working on right now. But it's good to have confirmation that we are on the right track.*
Ideas whose benefit you are not entirely convinced about	*Is that something the rest of you would also like to have?*
Ideas that are clearly absurd	*OK that sounds interesting. I'll suggest it to my team and we'll see what we can come up with.*

After their feedback, it may be useful to ask a follow up question:

Have you any idea about how that might be implemented?

Where exactly would you like to see that feature?

So how do you do that at the moment?

11.19 Promote an atmosphere where everyone feels relaxed about giving feedback

One common problem is that some people will give you a lot of feedback, while others will give you very little. The reasons for this may be that some people:

- are much more confident than others or want to make a good impression to someone else in the audience

- are afraid of being judged or are simply very shy

You need to promote a relaxed atmosphere in which everyone feels free to give feedback without any sense of judgment. It doesn't matter if the boss is there, he / she will appreciate their effort even if sometimes the ideas are not particularly pertinent or useful.

If you have someone who always try to dominate the discussion, you can say:

Do you mind if I just interrupt you a second, as I really wanted to hear what Pietro has to say on this.

Could I come back to you on that as I was wondering what the others thought about it.

On the other hand with shy participants a good solution is to give them eye contact and address them by name:

Stefan, what has your experience been of this?

Miranda, I know you have encountered this problem before, how did you resolve it?

However, be sensitive, and if it becomes clear that they really do not wish to participate, then perhaps it is best not to insist on asking shy participants questions.

11.20 Tell the audience when you are 10 minutes from the end

Ten minutes or so before the end announce that there are 10 minutes left. Look at your watch and say:

OK, we're very close to the end now, but there are just a couple of important things that I still want to tell you.

This will

- wake people up and enable you to regain the audience's attention at a very important moment in the presentation

- give them an opportunity to formulate any questions they may still want to ask

11.21 Exploit your concluding slide to leave a positive last impression

Your audience's final and last impression will probably be directly influenced by how you end the presentation.

Your final slide must be dynamic and totally audience-oriented.

First, stand confidently, look directly at your audience and pause for two seconds. This will signal to the audience that you are coming to an end. This is important as it will wake them up and get them to concentrate on your final points.

> Well that brings me to the end of the presentation. So, just to recap: [you then show your final slide that summarizes the three main points]

Second, end your presentation with some firm conclusions, preferably the three main benefits to them of what you have been saying. Answer the question 'now what?'

Compare these two versions:

STANDARD IMPROVISED ENDING	DYNAMIC SCRIPTED ENDING
So just to recap some of the main points I have presented in this presentation. X will give you a business-level interface compared to the technical interface provided by the current Y. Another thing, a greater level of standardization will allow you to have the possibility of more stability and will enable you to interface more and more clients with a single piece of software. Oh yes, and there will be a considerable reduction in the work time needed to carry out client interface tasks.	OK, just to recap. You are going to get three main benefits from X:
	First, a truly business-level interface.
	Second, you'll have more stability and you'll be able to interface with more and more clients with just one software application.
	Third, you will spend around 30% less time having to interface with clients.

The example above highlights the importance of knowing exactly what you want to say at the end of your demo. You can prepare such an ending by writing down every word (see 4.19) and then practising it.

11.22   Before saying goodbye give any further details

The last things to say

1. thank the audience

2. ask them if they have any questions

3. tell them where they can find the relevant documentation, handouts etc

4. tell them whether they can / should contact you (give your details) or someone else

5. thank them again (this signals to the audience that they can get up and go!)

12 CONDUCTING A DEMO / TRAINING VIA AN AUDIO CONFERENCE CALL

12.1 Be aware of the typical problems

In a typical audio conference

- the presenter is located at his / her company and the participants at their company

- the presenter and the participants cannot see each other, but they can see the same file on their monitor

- participants can send messages to the presenter / trainer and vice versa

- there may be background noise or echo

Because you cannot see the participants, you cannot:

- see how many participants are present

- establish any immediate relationship with the participants (e.g. there is no handshaking, no opportunities for chatting by the coffee machine)

- see / gauge the reaction of the participants, i.e. you are not helped by being able to interpret facial reactions or negative body language

- see if your participants are following your demo or are even watching it

- create much variety in the presentation style – the audience can only see the monitor, you cannot use a whiteboard or write things down on a piece of paper, or show them physical objects

- you can't use your hands or fingers to point to elements in your slides

The disadvantages for the attendees are that they:

- cannot see you or your body language (e.g. they cannot see if you are nodding your head in agreement with what they are saying)

- cannot connect with you on a human level and you risk being an anonymous voice

- cannot use their own body language to show that they are having trouble understanding you or that they want to interrupt you

- are in a very passive role and may find it very hard to concentrate on listening to you for more than 10–15 minutes at a time

A. Wallwork, *Presentations, Demos, and Training Sessions,*
Guides to Professional English, DOI 10.1007/978-1-4939-0644-4_12,
© Springer Science+Business Media New York 2014

12.1 Be aware of the typical problems (cont.)

Moreover, you may find it difficult to know if and when to ask participants questions.

The result is that often demos done via audio conference are much less effective than those done face-to-face or via video conference. Unfortuntately, in an audio conference the presenter often does his / her demo with very little interaction with the attendees, and with very little consideration of whether the attendees are following and can understand.

You have a duty not only to the attendees but also to your company to check that the attendees follow the demo. If the demo is not successful and the client fails to understand how to use the product or service, then

- the image of your company will be damaged

- your helpdesk will be inundated with requests for clarifications on how the product / service works

- the client may become frustrated and eventually decide to change providers

- you will not be satisfied with your work and you will be less motivated to do future demos

- if the call is with colleagues, they too will become demotivated and frustrated

Clearly, it pays to find ways of making demos via audio conference more effective. The rest of this chapter outlines ways to conduct better demos via audio.

12.2 Consider not using audio conference calls for non-interactive demos

If your task is to tell colleagues about a new project, product, service or system within the company, then a conference call is probably not the best way to do it. The problem is that your attendees will have a very passive role, i.e. they are purely recipients of information. If the call is not interactive, attendees will soon lose concentration and become demotivated.

Imagine if you had been taught mathematics over the telephone with someone you had never met and who talked to you for two hours at a time without you doing anything other than listening – how much do you think would learn and remember?

There is also a cultural problem. Some cultures, particularly in Asia, are not used to interrupting the speaker to ask questions or when they don't understand. So you may find yourself just talking and talking, and getting very little or no feedback from them. In such cases, it makes little sense to use a conference call to give such colleagues information.

A conference call, or any kind of meeting, is a costly event – you are taking several people away from their desks and their other work. If this is compounded by the fact that the participants get very little real benefit from participating in the call, then the reason for using a conference call loses most, if not all, its value.

In summary, the only way a conference call will work is if the attendees actively participate by

- asking questions and answering questions (see Chapter 13)
- carrying our frequent tasks (see Chapter 14)

If such participation is not possible, then don't use a conference call, much better is a video conference. If a video conference is not possible, two alternatives are:

- provide participants with a written document which they can then read when it suits them, and which they can annotate with their questions and comments
- make a video of yourself doing the presentation. The 'attendees' can then watch it in their own time, and have the opportunity to repeat parts that they didn't understand, and to watch it over several sessions

12.2 Consider not using audio conference calls for non-interactive demos (cont.)

Both these solutions enable you to provide the same presentation to several groups of people, without actually having to be physically present.

When participants have read the document and watched the video, your role is to:

- check that people have read the document and / or watched the video

- answer any questions they may have

You can do this either by email or by organizing a short audio-conference call. Another advantage is that people will not feel shy or embarrassed about asking questions because they will not be doing so in front of their colleagues, but only directly to you via email.

12.3   Exploit the advantages of audio calls over face-to-face presentations

Doing a demo via an audio conference call is not all bad news. In an audio call, the participants cannot see you. They do not know, for instance:

- how you are dressed
- how nervous you look
- whether you are consulting your notes
- whether you have a secret assistant who is helping you
- whether you are sending an urgent mail or text message at the same time as talking to them

You can use the fact that the demo is being conducted remotely by scheduling longer breaks during which you can:

- prepare answers to their questions
- go over the next sections / steps of your demo
- relax a little

12.4   Find out about the audience and have the mailing list of those attending

You will give a much better performance if you have a mental picture of the attendees (see 3.5 to learn how). You may find that if you have seen a photograph of someone it will make it easier to relate to them over the phone.

As with face-to-face demos and presentations, if you know who your audience is in advance you can decide how technical to be (will just technicians be present, or also business managers?).

You also need to tell them what you expect them to know (e.g. familiarity with certain software applications, knowledge of similar products) in order for them to get the most from your demo.

Given the problems of not being able to hear clearly during an audio call, it is a good idea to have the possibility to email people with documents or answers to questions. Likewise, you can ask them to email questions to you – even during the course of the demo. So ideally all parties need to have a messaging system available.

12.5   If possible, have a member of your team on site at the client's offices

If you know that one of your colleagues happens to be at the participants' workplace then try and enlist his / her support for your demo. This colleague will be able to understand better than you if the attendees are following the demo, and he / she can also answer their questions after the demo has finished.

12.6   Ensure that everything is OK from a technical point of view

If you are the presenter, then it may be your job to take care of the technical side. This entails:

- ringing telephone numbers given by the audio conference software application

- ensuring that your screen is shared with all the attendees

- making sure the sound quality is acceptable (you may also hear background offices noises from where the various attendees are located)

- checking that people can hear you well

- checking that everyone is there (generally a pre-recorded voice automatically announces when a new person has joined the call, and their name is added to a window containing the list of people who have currently joined)

- ensuring that everyone can see the screen and your first slide (document, figure etc)

12.7 Begin the call with some introductions

Don't launch immediately into the demo. First, introduce yourself. Tell the attendees:

- your name

- your position and how long you have been with the company

- why you have been chosen to do this demo i.e. how the work you have done and the expertise you have gained means that you are the right person to do the demo

- how and why you understand their work situation, i.e. you want to try to empathize with your audience and show them that in some respects you are very like them.

For example you could say:

My name is Leo Tolstoy and I have been in name of company *for three years. I am in the xxx group, which is why I have been chosen to do this demo. I actually helped develop the product that I am going to present to you. This means that I am very familiar with most of its features. Given that I once had a very similar job to yours, I hope to be able to give you some great tips and insights. And don't worry, we can take as much or as little time as you need.*

This mini introduction allows attendees to:

- switch their brains from what they were doing two minutes ago to the new activity (i.e. following your demo)

- tune in to your voice without having to concentrate on vital information

- feel that you are on their side

It is vital that you prepare this introduction in advance and that you write down every word (to understand why see 4.5).

Then ask attendees to introduce themselves. Ask them to tell you:

1. their name

2. why they are participating in the call

3. what they hope to learn from the demo

The answers to the above three questions will help you to check who is present and why, and tune into their voices.

These introductions allow everyone to become a little more relaxed and thus mentally prepared for the demo.

12.7   Begin the call with some introductions (cont.)

Then:

- ask them if they have done an audio conference call before, and how the experience went – then reassure them that this call with you will go much better!

- ask them if they know why they are on the call (their boss may have just told them to join the call but without explaining why) and what they hope to gain from the call

- reassure the audience that you will speak slowly and clearly, and that you will be very happy to hear their questions, and that in fact, the whole point of the call is to hear their questions and opinions

The result should be that they are less worried about the dynamics of the call (e.g. whether they will be able to understand, if and when they can ask questions), that they feel they have a purpose for being part of the call. Without this preparatory work, your call may become an extremely tedious time-wasting and unwelcome process for all participants (including you!).

12.8 Set some ground rules

Your attendees need to know from the start what to expect:

- what files should they have uploaded and have in front of them?

- are there any support files that they need to print so that they can refer to them during the demo?

- how long will the demo be?

- will there be any breaks?

- how will it be structured?

- will there be any hands-on tasks?

- when can they ask questions?

- what should they do if they can't understand you (i.e. your English not just the content)?

- do they need to follow all the demo? or are there some parts which they could skip?

- will the demo be made available on your company's website?

Clearly the best option is to deal with such questions in a preliminary email. But you cannot guarantee that participants will actually read your email. So you should prepare in advance a short speech that answers all the above questions. For instance you could say:

Before we start / Now that we are all here, I would like to check a few things with you. First, you should all have a copy of the demo on your screens. Has everyone got it uploaded? Secondly, it would be useful if you could have a copy of the manual, as I will be referring to it occasionally during the demo. If you don't have one, I can email it to you straightaway.

OK, as I mentioned in my email to you, the demo should take around two hours. I plan to have a ten-minute break in the middle. Does that sound OK? I have structured the demo so that we alternate between theory and practice. So there will be plenty of opportunities for you to test out the features.

I hope that you will have lots of questions to ask. But I think that given that this is an audio call it would be best if you did not ask questions all the time. So I will make sure that I give you opportunities to ask questions at the end of each theoretical section. But if I forget, please remind me. I will also try to make frequent summaries and this will be another opportunity for you to ask questions.

When you do ask a question, could you say your name first and then ask the question. Thanks.

12.8   Set some ground rules (cont.)

As you can tell from my accent, English is not my first language. I will try to speak slowly and clearly. If occasionally you can't understand something because of my pronunciation, then feel free to interrupt me.

OK, please could you now look at the second slide. The agenda. Could you quickly look through it and tell me if there are any parts that you think I don't need to cover. For example, are you all familiar with the feature mentioned in the third bullet? OK. So we can leave that one out.

Finally, there is a full version of this demo along with additional notes on the website. I sent you the address in the email last week. By the way, do I have everyone's email address?

OK so that concludes my introduction. Any questions?

A big advantage of the audience not being able to see you is that you can prepare written speeches for various parts of the presentation, which you can simply read out or use as a prompt.

12.9 Don't use exactly the same style as you would in a video or face-to-face demo

Because you can't see your audience, you have no visual clues to help you understand if the participants are following you or not. You cannot see the expression on their faces, you cannot see them leaning forward because they are unsure of something, and they cannot raise their hands to interrupt you.

This means that you have to adapt your usual face-to-face style in order to account for these problems. You can partially compensate for lack of video by:

- speaking more slowly than usual

- focusing less on theory and more on practical examples that the audience can easily relate to. In any case, they can learn more about the theory from a written document that you could email them

- emphasizing key words through your tone of voice

- making transitions in your presentation very clear not just through the change of slide but also with your voice

- using the pointer on your mouse slowly and effectively

But most importantly you need to interact with the participants more frequently. You can do this by asking questions and getting them to do more tasks (see next chapter).

12.10 Number your slides and constantly remind the audience where you are

During an audio conference call, attendees will usually be able to share a screen with you on their monitor and follow your slides. But they might only have a printed version. So when you move from slide to slide it helps if you say *OK, let's move on to slide 16* rather than saying *let's move on to the next slide*. Also people may leave their desk during your presentation, so when they come back it helps if they know which slide number you are up to. Thus it is also important to number each slide.

12.11 Make sure your demo is interactive and not robotic

If your audio conference call is for a long session (e.g. 1–2 hours) you may find yourself talking like a robot and almost forgetting that people are listening to you. So if you are getting no signs of reaction from your audience, then it could be that your attendees:

- are reluctant to interrupt you for cultural reasons

- are completely lost

- have stopped listening because they have failed to see a reason for listening

- have lost concentration – we don't normally sit passively for long periods of time with no visual clues

Given that you can't see their faces, you feel tempted to continue talking until the call finally ends. But if you do this, your attendees will feel very frustrated and demotivated, particularly if you have scheduled other calls with them to continue with your presentation.

It is thus essential to find ways of making the call more interactive and of adding variety. These include:

- having pauses every 20 minutes, where participants have to email / sms you two or three questions that they would like answered

- asking specific attendees to make summaries (verbal, or again via email) of what they understand so far

- asking them to tell you what they think your top three key points are so far

- pausing yourself to make summaries for them

- constantly highlighting why it is important for them to listen, and the benefits of doing so

Using email for participants to ask questions is perfect for those attendees who are reluctant to talk because they:

- don't want to lose face in front of colleagues

- are too shy

- think their English is not good enough

12.11 Make sure your demo is interactive and not robotic (cont.)

Simply asking:

OK?

So far so good?

Do you all understand?

is not enough – participants may simply say 'yes' (even if they have understood nothing) or may not even respond at all. Also, you need to give them time to formulate their questions. You need to wait several seconds, before you decide to continue with the presentation.

To learn about other questions you can ask you attendees see Chapter 13.

12.12   Learn how to deal with on-the-spot oral questions

The added difficulty of answering questions over the telephone is that the sound quality is worse (and of course you have no visual clues). But you can use poor sound quality to your advantage by saying that the reason you cannot understand a question is because *the line is bad* rather than that you have no idea what the question is!

One typical danger is that you will only hear a few words of the question, and you will model your answer to react to those few words. The problem is that the few words you hear may not actually be the most relevant part of the question.

So, if you don't understand the question – whether due to real sound quality problems or not – you can say:

> *Sorry but the line has gone bad and I can't really hear the question. Do you think you could email it to me, then I will answer it straight away on the phone? Thanks.*

To learn about how to understand questions from participants see 15.11.

12.13   Announce clearly that you are about to conclude the call

In a face-to-face presentation it is normally clear to the audience when the presenter has finished. He / she may sit down, stand up or make some other visual clue to indicate that the demo is over. In an audio call you need to make it absolutely clear that you have finished, for example by saying:

> *OK. Unless anyone has any further questions I have concluded the demo. As I mentioned at the beginning, the demo is available on the website. Also, you have my email address, so don't hesitate to contact me. I may answer your mail myself, or direct it to the relevant department. Pause OK. Thanks very much. I hope it was useful. Goodbye.*

12.14 After each session, revise your demo to improve it

Whatever type of presentation you do, it always pays to write down a few notes on how it went immediately after you have finished. These notes should include:

- improving how you set up and managed the call
- reducing the length of the call / demo
- making the slides more readable and effective
- asking more effective questions
- involving the audience

12.15 Get feedback on your performance

One of the best ways to improve your performance is to get feedback from the participants. You could ask for feedback directly via email, however participants are unlikely to answer, and if they do, they are unlikely to be very critical of you.

It makes sense to get someone else in your company to ask for feedback (this should actually be part of company policy). Participants will feel that this is a more official method. They are thus more likely to take it seriously and to provide useful comments.

13 GENERATING QUESTIONS DURING A DEMO / TRAINING SESSION

Note: Unless otherwise specified, the subsections refer both to face-to-face demos (including video conferences) and audio conferences.

13.1 Understand the importance of asking your participants questions

The success of your demo depends on how much your audience understand. There are two main ways to check if they have understood:

- by setting them a practical task to do which is related to what you have just explained. Their level of success in this task will be a clear indicator of whether they have understood or not

- by asking them pertinent questions

Asking your audience questions during training is fundamental. You need to be absolutely sure they understand. Resolving problems now means avoiding problems (and costs to the company) in the future. If you don't check how much your audience have understood, there is a risk that they will be forced to ask your helpdesk questions which you could have easily answered yourself.

If you don't ask any questions, the risk is that you will be the only person talking and this will be extremely tiring for your participants.

An additional problem is that people of whatever culture:

- are reluctant to admit that they haven't understood, particularly as they erroneously conclude that the rest of the audience will have understood

- do not ask questions because they fear that they may not be relevant for other members of the audience, or because they are embarrassed about their level of English

The result is that only those with good English will ask the questions. And those with the low English will probably not even understand the questions asked by the good-English-speaking colleagues.

A. Wallwork, *Presentations, Demos, and Training Sessions,*
Guides to Professional English, DOI 10.1007/978-1-4939-0644-4_13,
© Springer Science+Business Media New York 2014

13.1 Understand the importance of asking your participants questions (cont.)

This chapter deals specifically with questions in audio and video conference calls where a demo or training session are being given. However, you should also read Chapter 15 on dealing with questions in more formal presentations, as much of that chapter is also relevant here.

13.2   Don't just ask 'OK?' to check understanding

You need to ask the right kind of questions. The questions below are generally <u>not</u> very effective – they are the equivalent of just saying *OK?*

Is that clear?

Does everyone understand?

Everybody with me?

Does anyone have any questions at this point?

Does that all seem to make sense?

There is a high risk that the participants are too embarrassed to admit that they don't understand. The easiest solution for you is then to say to yourself *OK, they understood* and then move on to the next point. However, if they have not understood much of what you have explained so far, they will probably understand even less in the rest of the demo. You risk losing your audience completely.

So you could ask a question that is directed to all the participants but which requires them to answer. For example:

I am worried that I may not have been clear – would one of you mind just telling me what you've understood about this?

But again, everyone might simply look down or away, and you are left in embarrassing silence. In any case they could quickly get stressed out if they think you are going to ask them a question. Instead you need to make the question format less threatening (see next subsection).

13.3 Ask direct but non-threatening questions (face-to-face / video)

First, choose someone who is giving you eye contact – not those that are looking down or look as if they are panicking. If they are giving you eye contact it probably means that they know the answer. They then answer the question and hopefully those that did not understand now have a chance to hear the same things expressed in a different way by one of their colleagues. Then you can ask someone else (maybe one of those who was looking down before) if they have anything to add or any questions to ask.

Don't always direct the questions to the same people. Ask easy questions to participants who seem to be following you less. If your question is easy to answer this will provide a positive experience and will make such participants feel more relaxed and more likely to ask questions later.

In a face-to-face demo you can choose a participant and ask them a direct question such as:

Can you just explain what you have understood to be the main function of x.

What part of my explanation of x did you find most difficult to understand?

The worst participants to choose are those whose body language clearly indicates that they will be unable or unwilling to answer the question – for instance, those who are deliberately not giving you eye contact or who look anxious. In an audio demo you don't have this visual information.

Instead you need to ask less threatening / demanding questions. But in any case you can choose a specific attendee:

David, for you which is the most important component and why?

Sara, are there any features that you think that you will probably not use?

Silvia, which of the last few slides do you think that you or any of your colleagues would like me to go over again?

Mike, how do you think you personally could use this feature?

Peter, what would be the benefits of x for you?

Karen, is there anything about the product that you don't particularly like?

The above questions are less threatening because they are apparently not checking the participant's understanding but simply asking their opinion. If you simply ask *Is that clear?* you are unlikely to get the same level of detail as you will with the above questions.

13.3   Ask direct but non-threatening questions (face-to-face / video) (cont.)

If participants know that they are likely to be asked questions, this will also help them to focus more and be more attentive. They are also more likely to retain the information that you are giving them. On the other hand, if they just listen passively they will

- lose focus and interest

- remember less

- have more need of your helpdesk at a later date

13.4 Use your body language to encourage the audience to ask you questions (face-to-face / video)

You can encourage people to ask questions if you move closer to the audience. On the other hand if you move backwards in a rather defensive position with your arms crossed, the audience will have the impression that you do not actually want them to ask you questions because you are not really interested or that you are worried you will have problems.

13.5 Choose specific people to answer questions (audio calls)

When you cannot see your audience, you obviously cannot direct a question at an individual simply by looking at them.

Your tendency will probably be just to ask the question to no one in particular and hope that someone will answer. Possible results of this approach are:

- if there is a hierarchy within the audience, then the audience may wait for their boss to answer the question (this is fairly typical behavior in some Asian countries)

- the same person / s might keep answering the question (you may not even be aware of this as you may not be able to distinguish between the voice of one person and another)

One solution is to ask no one in particular the first few questions. If it is apparent that the same people keep answering the questions, then you need to start asking individuals. You can choose individuals from your list of participants.

In any case, asking individuals is a much better solution:

- you are almost guaranteed to get an answer (the person cannot remain silent)

- it will keep the entire audience attentive as they will not know who will be asked next

13.6 Tailor your questions to particular people

You will get much more useful answers to your questions if it appears that your question has been specifically designed for a certain person or set of people.

For example, when addressing a question to a group you can say:

> *So, **I know that many of you have had problems with the new system,** would you just like to tell me how it has affected you personally?*

To an individual:

> *Richard,**on your feedback form for the last session you mentioned x and y.** Could you explain what you said to the rest of the group, as I think it is very pertinent to them too.*

By beginning your question with some preliminary statement about the specific individual (i.e. the parts in bold in the two examples above) you show that

- you have taken the trouble to find out about your audience
- you are really interested in hearing their answer

13.7 Say something positive about the individual to whom you are addressing the question

If you say something positive (and genuine) about the person then this will further motivate him / her to answer your question. The positive statement is in bold in the examples below:

> *Richard, on your feedback form for the last session you mentioned x and y. Could you explain what you said to the rest of the group, **as I think it is very pertinent to them to.***

> *Carlos, **I know that you have made a major contribution to the success of the project,** so perhaps you could tell us why.*

> *Helen, **I thought what you were saying before the break was absolutely fundamental.** I wonder whether you could expand a little more on it.*

13.8 Give attendees a chance to prepare their answers

If you have emailed your presentation in advance, then you can tell the attendees in your email that you would like, if possible, for them to prepare answers to any questions contained on your slides (see 13.11).

If you have not emailed them in advance, and thus they are seeing the 'questions slide' for the first time, then you can say:

OK, so here's a list of questions that I would like you to answer. So we'll have a pause for five minutes, while you formulate your answers.

This is then a perfect opportunity for you to stop talking, for them to have a break from listening to you, and for everyone to relax for a moment. I cannot stress how important it is to integrate / schedule such breaks into your presentation.

You could instruct them to do one or more of the following:

Can you write down the three most important things you have learned so far.

Write down two things that you are not totally clear about.

Of the features I have shown you so far, which do you think would be the most useful?

And the least useful?

There are many other advantages to this idea:

- you can choose specific participants to read out their answers. This means that you can encourage the quieter participants to speak. Generally speaking this should work fine as the participants will have already prepared their answers so they should be less embarrassed or anxious about speaking. But avoid this technique in an audio call, as you will not be able to see from their face whether they are reluctant to speak

- you will understand what your audience perceive as the benefits and downsides of your company's products and services. They may not have fully understood what the benefits are, so you then have a chance to underline such benefits

- you have the perfect opportunity to go over the difficult points again

If your demo is over several days, then the above technique is also a good means for you to check understanding and to summarize the activities of the previous day / session.

13.9   Get participants to explain things to each other

Sometimes you will ask a yes / no question i.e. a question that can be answered simply by saying 'yes' or 'no'. These questions are usually best avoided as they do not provide you with much information.

But occasionally you may decide to use them. If you do, there is a good chance that one or two people will say 'yes' and the rest will say nothing. You can probably assume that the others don't know the answer.

In such cases follow this two-part strategy.

First, ask the participant who said 'yes' to explain to the others what he / she thinks.

OK, so Stefan you know the answer. Would you mind just explaining to the others?

Second, when this participant has given their explanation you can:

- either confirm or adjust what they said

- or ask another participant to comment on the first participant's explanation, to check whether they agree with each other

The benefits of this strategy are:

- those participants who didn't know the answer to your question will now hear the answer from a colleague whose English they may understand better than yours

- you have an opportunity to check whether the explanation is correct or not

- the focus is off you and thus the presentation becomes more interactive, not just between you and them, but also among the participants themselves

13.10 Schedule a break for dealing with participants' questions

One advantage of doing remote demos is that your audience are at their own office. This means that you can have long breaks in your demo, where they can return to their desk and carry out their normal work. You can exploit such breaks for dealing with their questions.

Let's imagine that you have already done one hour of your demo. You feel that although some participants have asked questions, there have been some participants who have remained silent throughout but may have questions to ask. So you can say:

> As we agreed at the beginning of this call, we are now going to have an hour's break. Could I ask you a favor? Do you think you could each email me a few questions that you would like me to answer? Then during the break I can look at the questions, and decide how best to answer them. My email address is on the slide that I am putting up now. OK? Right see you in an hour, at 11.15 to be precise.

This is a great strategy because it means that you have:

- the opportunity to assess how successful your demo has been so far, i.e. how much the audience has really understood

- a much greater chance of understanding the question if you receive it via email rather than having to decipher it over a telephone line

- time to formulate your answers

When you restart the demo after the break you can say:

> Thanks very much for your questions. There are one or two that I think would be easier answered via email, which I promise to do by the end of tomorrow. There are also a couple that I think are best dealt with now. So let's look at them together. You should be able to see them on your screen now.

It is best to cut and paste their questions into a slide so that everyone can see them clearly. This also means that you can ask them to read the question, and then you can answer it.

13.11 Incorporate questions into your slides

A good way to make your presentation interactive is to have specific slides with questions on. Having questions provides variety – you can have a break from talking and your audience can have a chance to interact. It also puts your listeners at the center of attention. If they know they will be asked questions, they are much more likely to try to follow what you are saying and participate actively.

So when should you have 'question slides'? A good time is after you have explained a concept over several slides. These questions can do one or more of the following:

• test how much the participants have understood

• get feedback on their level of interest in the topic

• find out more about how much they know about the topic

• find out what further details they would like to know about the topic

Let's imagine you are training a group of young salespeople on how to write emails to potential clients. You have just given them some ways to write effective subject lines. You could then have a slide that looks like this.

Subject lines: Questions

1) Think of the typical subject lines that you write – which one has been the most effective? Which ones don't seem to be effective at all?

2) What unusual subject lines have you received that have made you want to open the email?

3) What kind of subject lines make you want to trash the email without reading it?

The idea is not just that you want to hear their answers. By asking questions at regular intervals during your presentation, you ensure that the audience will participate much more and that the presentation will be more dynamic. If the focus is only on you, the audience will quickly lose interest, and because you cannot actually see the audience, you will not know whether they are listening or not. By asking them questions they are forced to listen to you.

Here is another example. In this case the presenter shows the participants a slide with a problem to solve:

FALL IN SALES

The number of repeat orders has gone down 12 % over the last year.

Findings from customer surveys do not reveal any increased levels of dissatisfaction.

What could be the reason for this discrepancy?

13.11   Incorporate questions into your slides (cont.)

Below is the same information as in the above slide, but formatted slightly differently to distinguish the problem from the question.

FALL IN SALES

- The number of repeat orders has gone down 12 % over the last year.

- Findings from customer surveys do not reveal any increased levels of dissatisfaction.

Question: What could be the reason for this discrepancy?

Basically, if your slide is designed to get the audience thinking about an answer to a question, then the question needs to stand out very clearly.

13.12 Decide the best way to phrase your question slides

As with all questions that you ask you need to decide what kinds of questions will work best. To do this you need to ask yourself:

- how will they interpret my question? Is there only one possible interpretation?

- how are they likely to answer this question? Will their answer be useful to me?

- how focused is the question? Is it so generic that the answer will be too generic? Or is it so specific that it might only be relevant to one person in the group?

- at this point in my presentation is this a good question to ask? Might the question be better later (or earlier) in the presentation?

You may not be the best judge of your questions, so it is worth asking a colleague to have a look at them.

If you have a question slide with three or more questions you can then ask the participants to choose which questions they would most like to answer. This gives them a feeling of power and also means that they will choose the question that they feel is the easiest for them to answer or which is most pertinent to them.

13.13 Have recap slides in addition to or as an occasional alternative to questions slides

You can create variety in your demo if you occasionally insert slides containing a summary (i.e. a recap slide). You can ask your audience to read the summary and ask them which points they would like you to explain again, or which points they found the most difficult to understand. You can then say

Can you explain exactly what it is that you did not understand?

This should then automatically force them to ask you for clarifications.

13.14   Go over any key points more than once

Learn to predict what elements in your demo some participants are unlikely to understand. For example, it may be one bullet point in a series of five. If you ask if they understood all the points, most will nod. Those who do not nod may however be reluctant to ask because they feel that it undermines their credibility in front of their colleagues. So, if you see that someone is not nodding and you think it is an important point, then you can say:

Bill, which point would you like me to explain again?

In any case, I think it is worth going over some of these points again.

Given the importance of this point, let's just look at it again.

13.15   Signal to the participant that you are listening to and following his / her question

When you are face to face with someone who is answering your question you can show that you are following their answer by nodding your head. Clearly, this does not work in a purely audio conference. So remember to make noises to show that you are following, e.g. *OK, ah, yes, I see.*

You can also make the process more interactive by intervening in their answer, for example by saying:

So you mean that ...

So what you are saying is ...

Right, so you think it is best to ...

So, you would actually prefer to ...

What you are doing is paraphrasing what they are saying or interpreting it. This allows you to do two things:

• show the attendee that you are interested in what they are saying

• clarify / check that you have understood correctly

Don't simply remain silent.

14 MAKING YOUR DEMO / TRAINING INTERACTIVE: SETTING TASKS

14.1 Always mix theory with practice

Imagine that in your English lessons at school you had never done any exercises or never talked, but simply listened to the teacher giving explanations the whole time. How much would you have learned? The Chinese have a proverb: *I hear and I forget, I see and I remember, I do and I understand.*

If you are doing a training session, then obviously the best way for the audience to learn and understand, is by putting into practice what you have just taught them. So prepare exercises for them to do. But don't divide the training session into two parts – one theory and one practice. Instead:

- alternate theory with practice

- occasionally do the practice first and then explain the theory

- or do the practice and then get them to explain the theory – if they discover it by themselves they will retain it much better.

A. Wallwork, *Presentations, Demos, and Training Sessions,*
Guides to Professional English, DOI 10.1007/978-1-4939-0644-4_14,
© Springer Science+Business Media New York 2014

14.2   Decide the best tasks to give participants

The kinds of tasks you can give participants include getting them to:

- write down questions
- write a summary of what they have understood so far and how it will be useful in their jobs
- explain in their own words what you have done so far
- test out the product (i.e. a software application, a physical device)
- follow a list of instructions whose correct completion will signal to you that participants have managed to follow what you have explained so far
- solve a problem

All of above can be used with any kind of group. On the other hand, the following will only work in a face-to-face situation:

- produce a diagram summarizing a particular point
- get up and explain one of the slides from the earlier part of the presentation
- talk to each other to discuss or resolve some issue
- recall a particular situation in their own lives, or the life of their company or country

14.3   Give clear instructions when setting tasks

Setting tasks requires preparation on your part, not only in terms of selecting an appropriate task, but also in describing to participants exactly what they have to do. It is a good idea to test the task first with some colleagues. This will reveal a) whether your instructions were clear b) how difficult it is to complete the task c) how long the task takes.

It is easiest if you give participants written tasks to follow. Do not explain the task in a series of one or more paragraphs. Instead:

1. divide the task into clear steps
2. put each step into a separate numbered paragraph
3. if appropriate, indicate how long the task will take

Basically, the task should be set out as in an instruction manual. To learn more on this, see the companion volume *User Guides, Manuals and Technical Writing*.

14.4 Don't read aloud the task outlined in your slides

You may decide to write the task onto a slide and incorporate it into your demo presentation. The task below is a problem to solve (rather than a series of steps to follow) so the bullets are not numbered.

DEALING WITH CUSTOMER COMPLAINTS

- The help desk has informed you of a customer's problem.
- The customer says that ...
- The customer has also provided evidence of ...

Question : What should you do?

In such cases, do not read out the example. Instead say *please can you read the example on slide number 10 about dealing with customer complaints*. Pause while they read – a good amount of time to pause is to read the slide to yourself slowly and twice. This should give the audience enough time to absorb the information. Then choose a participant and say: *So, Zach, what would you do in this situation?*

As highlighted in the above slide, examples are best combined with a question that you want your audience to answer.

14.5 Use the whiteboard or a new window on the shared desktop

When you are explaining something that is not clear, in a face-to-face demo you can stand up and use the whiteboard. This has the advantage of attracting attention as well. In an audio call you could open a new text window on the shared desktop, and use this as a whiteboard.

14.6 Don't wait for everyone to finish the task

In a face-to-face demo, you can go around and monitor everyone's progress. When around two thirds of the participants have completed the task, then tell everyone to stop. If you don't, you will waste time waiting for the slow people to finish.

In an audio call, tell people to let you know when they have finished.

If you think it is important for everyone to finish the task, then schedule a break for the end of the task so that those who have finished can go off and do something else, but those who haven't can complete the task.

14.7 Give participants positive feedback

Ensure that when you give feedback, you don't merely say OK or nod your head. Try to say something positive and encouraging.

Finally, explain why it was important for them to have done the task, and what they have learned from it.

14.8 Avoid distractions: Get attendees to share a PC

If your demo involves participants using a PC, get two or more participants to share the same PC. If you don't there is a risk that they will start googling, emailing, facebooking etc rather than concentrating on you or on what they are supposed to be doing.

14.9 Ensure that every training session / demo is interactive and contains a practise session

It doesn't matter how technical your demo is, there are always opportunities for practise. For example, imagine you are explaining to the audience the new features that have been added an old product. Your aim may be to get a client to buy the new version of the product, or you may be teaching colleagues in your company to use the new product.

Clearly you could get the audience to try out the new version. But you can do much more. To motivate them to try out the new version and thus to see how effective it is with respect to the previous version, get them to do the same task twice. First, they use the old version to do the task. Then they do it again with the new version. This will clearly highlight the benefits of the new version.

You can use the same trick to compare your product / service with that of a competitor.

14.10   Break up your demo with fun activities

Not all audiences have the same attention span. Some countries have rigid and quite conservative education systems such as Italy, India and Japan. In these countries lessons may be very teacher / professor-focused, with very little interaction with the students. Students are expected to listen to, memorize and recall vast amounts of information.

The result is that the expectations of people who have been educated in such a way are quite different from those people from Anglo countries (e.g. the UK, US, Australia, Canada, Australia, South Africa, Ireland, and NZ) or northern Europeans. In Anglo countries an audience are only likely to absorb about 20 % of the information you give them, and are also likely to forget 75 % of what they heard within 24 hours.

Studies of students educated either in Japan or in the US, have shown that the Japanese have far higher attention levels than typical north Americans, and are also much more persistent when carrying out a task.

Anglos and northern Europeans expect not just to be informed, but also to be entertained. They do not want to be subjected to the phenomenon known as 'death by Powerpoint'!

This means that you cannot deliver your training or presentation in a karaoke style, where you simply read aloud your slides to the audience (you might as well just email the slides to the participants and let them read them at their own leisure).

Your slides need to be easy to follow (see Chapters 5–8). But even if your slides are fantastic, there will come a point in the session when the audience starts to lose interest.

To stop them losing interest, every 15–20 minutes I suggest you give them a short fun activity. Such activities include:

- logic tests
- lateral thinking games
- interesting factoids

14.10 Break up your demo with fun activities (cont.)

Here are three examples of mathematical games, with quite counter intuitive answers (see end of the chapter for the key). Note: you are not allowed to use a paper and pen, or to make any written calculations. You just have to give a very quick answer.

> PROBLEM 1 You can't get to sleep. You decide to use the old-age method of counting sheep. How long would it take you to count (1, 2, 3,...) one billion sheep? Assume that it takes you one second to say each number.

> PROBLEM 2 You fold a (very big) sheet of paper 64 times. How thick does the final pile get?

> PROBLEM 3 Take a sphere of 1 mm diameter, made of iron. You have 1000 of them. What is the total weight?

These activities do not have to be directly related to the topic of your training. They are simply designed to act as a short break to regain your audience's attention, and for you to have a few moments of relaxation. In any case, it is essential that the activity is short (not more than 2 minutes) and is easy to explain. You can find hundreds of such games on the Internet.

Including such activities is not a sign of an unprofessional presentation. As Martin Chalfie, Nobel Prize Winner in Chemistry told me:

> A professional presentation can be both serious _and_ fun.

Being both informed AND entertained is an essential requirement for an Anglo audience - it is not an option.

Key to mathematical games

1. 32 years.

2. A sheet of paper is about 0.1 mm thick. The number of layers doubles at each folding. Therefore you end up with $2^{64} = 1.8*10^{19}$ layers, which make a thickness of $1.8*10^{12}$ km, which is about 10000 times the distance between the Earth and the Sun.

3. It is about 5 grams. You can pack all of them in a small cube with side 1 cm. In fact, $10*10*10 = 1000$. One cm^3 full of iron is about 7.2 grams, but there is space between the spheres.

Lesson to learn: For the human brain, 1000 and 1 billion are just "big numbers". Moreover, exponential growth, like in game 2, is out of our thinking (we think "linear", that is proportional). Normally we underestimate answers 1 and 2, while overestimate answer 3.

15 ANSWERING QUESTIONS

15.1 Prepare in advance for all possible questions

Questions from the audience

- should help you to expand on what you are trying to tell them

- make the audience feel involved

- are an opportunity to learn what audiences really want to hear. This will help you when you have to deliver the same presentation to a similar audience

It may seem that you have no control over the questions the audience might ask you. In reality you do have some control, as long as before the presentation you give yourself time to prepare answers to possible questions.

Practice your presentation in front of colleagues. Choose colleagues from another team who are not so familiar with your product and service. Get them to write down questions that they think the audience might ask. Choose the ones that you think are the most relevant, then prepare answers to them.

If you have thought of all the questions your audience are likely to ask, it will enable you to

- seem professional by being immediately able to answer a question

- stand a better chance of understanding (in terms of the words the questioner uses) such questions when they are asked

- prepare extra slides which answer such questions

- prepare yourself mentally for difficult questions from difficult people, so that you can deal with them calmly and politely

- decide which questions would be best answered by saying: *I think it is best if you email me your question, then I will be able to give you a comprehensive answer*

A. Wallwork, *Presentations, Demos, and Training Sessions,*
Guides to Professional English, DOI 10.1007/978-1-4939-0644-4_15,
© Springer Science+Business Media New York 2014

15.2 Include the topic of questions in your agenda

Make sure that when you have outlined your agenda, you tell the audience when they can ask questions i.e. at specific moments during the demo, at any time, before/after the breaks, only at the end.

For example you can say:

Feel free to ask me questions at any point during the demo.

I would prefer you to only ask me questions at the end of each part of the demo.

I have scheduled a 15-minute question and answer session at the end of this morning's presentation.

15.3 Repeat the questions

Before answering, repeat the question you have been asked so that:

* the rest of the audience can hear the question clearly – this is particularly true if the question comes from someone in the front row, as the back rows will not be able to hear it; or when the questioner has a very low voice or speaks very fast

* you can reformulate any badly expressed questions

* you have time to think about an answer

* the questioner can check that you have understood his/her question

In any case, give yourself two or three seconds to formulate your answer, before responding.

So if the questioner says: *I understand how the first feature works, but I don't really understand why the second feature might be useful* or *Could you just explain again what happens if the machine breaks down?* You can say:

So you are asking about the utility of the second feature. Well, ...

So, you want me to summarize what happens if the machine breaks down. OK ...

By beginning your phrase with *so*, you indicate to the audience that you going to repeat the question.

15.4 Only interrupt a questioner when strictly necessary

Most people don't appreciate being interrupted when they are asking a question. However, if they are clearly having difficulty in expressing themselves and you feel it would be right to help them, you could say *So you are asking me if ...* Basically you are interpreting what they want to say, and saying it in your own words for them.

If their question is taking a very long time to ask, you can say:

> *Sorry, I am not exactly sure what your question is. I think it might be best if you asked me during the break.*

> *Maybe you could write your question down and give it to me at the end of the session.*

If you realize that the question has limited interest for the rest of the audience, respectfully say to the questioner.

> *For me this is a fascinating topic, but I think it might be best if we discuss during the break. If that's OK with you. Now, does anyone else have any questions?*

If you think their question is probably irrelevant to the rest of the audience, you can address the audience and say:

> *Has anyone else had this experience?*

> *Is this something anyone else has encountered?*

Hopefully the audience will stay silent and you can offer to answer the question after the presentation.

Whether you interrupt will also depend on who is asking the question. If the person is high up in the client's hierarchy, you must show respect for his/her position, and not interrupt.

15.5 Involve all the audience, don't just give eye contact to the questioner

Answer not only the questioner but also the whole group. Keep eye contact most of the time with the whole audience, but keep going back to the questioner to check from their body language (e.g. nodding, positive smiling) that they are happy with your answer.

Where appropriate include the audience in your comments:

> *I imagine several of you have come across this problem.*

> *Is this something that others of you have had difficulty with?*

Unfortunately some people just seem to like asking questions without actually being too interested in the answer.

174

15.6 Help the audience to think of questions

It is not enough to simply ask 'Does anyone have any questions?'. Here are two options for encouraging questions:

- use a summary slide to remind the audience what you have covered and the things they may want clarifying
- arrange with a colleague that he/she will ask you a question

You could also get the session going yourself by mentioning a question that someone else has asked you. For example:

Maybe I could start myself with a question that someone asked me yesterday ...

People often ask me ...

If this fails to get the audience to ask questions of their own, make sure you still sound enthusiastic and that you have not been discouraged.

15.7 Learn how to recognize if someone wants to ask a question and how to respond

Be aware of the kind of body language people adopt when they want to ask a question (leaning forward, mouth open, one finger up).

15.8 Be concise

When answering a question it helps to be concise, particularly as you might otherwise forget what the original question was.

If the question only requires the answer *yes* or *no*, you can be suitably brief and move on to the next question.

Sometimes you will get two-part questions. It's generally the best option to choose the part of the question that is simplest to answer first. If you forget the other part of the question, you can ask the question to repeat his / her question. Alternatively, you can move on to another question, and after the presentation talk to the questioner in person.

There are some questions that you could discuss for hours, but the questioner is not asking you to tell them everything you know about the topic, but just what is relevant to now. If you are tempted to begin a long conversation with someone in the audience, offer to meet up later.

15.9   Never make a participant lose face

Very occasionally questioners in the audience seem to want to provoke us, and one natural tendency is to become defensive. You don't need to take any criticisms or objections personally. Simply say:

> *I think you have raised an interesting point and it would be great if we could discuss it after the demo.*

> *I was not aware of that. Perhaps you could tell me about it during the break.*

Be aware that some people just ask questions to demonstrate their own knowledge. In this case, you can say:

> *You are absolutely right. I didn't mention that point because it is quite technical / because there was no time. But it is covered in the manual.*

Be very careful of your body language. Presenters who fold their arms when answering questions may be perceived as being defensive.

Never directly contradict someone, don't make them feel bad or stupid.

Sometimes you may need to get a participant to clarify what they have said in answer to your question or when they are asking you a question. Try to phrase your comment or follow up question in a positive way. For example, if you say something like:

> *Can you explain <u>better</u> what you mean?*

you are implying that they are in some way incompetent as they could have done something 'better'. Instead, make it seem that is your responsibility for not understanding.

> *Sorry, I am not sure I have understood completely.*

If you don't agree with what they have said, then at least show that you appreciate what they have said:

> *I see what you mean but ...*

> *I appreciate what you are saying, but ...*

> *Sure, that certainly makes sense, but ...*

If possible say something positive. This will then make the attendee feel good, and will also make the other attendees feel relaxed, so that they too know that you will treat them well when you ask them a question. This will increase their willingness to participate actively.

15.9   Never make a participant lose face (cont.)

The same is true if you cannot understand what they are saying. Don't keep asking them to repeat the question, as this could make it look like they are stupid. Instead say:

I'm sorry guys but there's a lot of noise on the line and in my office. Do you think you could message your question to me.

Can you hear me OK because I am having some problems hearing you.

In both cases above you are making it seem that it is a generalized problem of sound quality which has nothing to do with the attendee who is answering your question. This helps to save everyone's face.

15.10 Remember that it is not necessarily your fault if you can't understand the question

Do not be embarrassed to say that your level of English is low. Tell the audience at the beginning of your presentation (or at the beginning of the question and answer session) you may have difficulty understanding their questions and that they should speak clearly and slowly. Other members of the audience, whose English level is also low, will thank you for doing this!

In any case, your ability to understand the questions depends not just on you. It is also the responsibility of the questioner to phrase and enunciate the question in a way that you will understand it.

So, if you don't understand a question, particularly from a native speaker of English, simply say:

I am sorry, but like many people in the audience, I am not a native English speaker. Could you speak a little more slowly please? Thank you.

Alternatively you could say:

Would you mind emailing me that question, and then I will get back to you?

Do you think you could ask me that question again during the coffee break?

Sorry, I really need to check with a colleague before being able answering that question.

15.11 Learn strategies to help you understand and clarify questions

If you don't understand a question, the best solution is probably <u>not</u> to say:

Could you repeat that please?

Sorry, I did not understand, could you say that again?

The problem is that the questioner may simply repeat the same question with the same words at the same speed.

If they have spoken too fast or with a very quiet voice, you can say:

Could you speak more slowly please.

Could you speak up a bit please.

If you still don't understand, then you can try to identify the part of the question that you did not understand. For example:

Could you repeat the first part of your question?

Sorry, I didn't catch the part in the middle / end.

Often the most effective solutions is to repeat back the part of the question that you understood until the point in which you stopped understanding, as indicated in the table.

	THEIR QUESTION	YOUR CLARIFICATION
1	Do I need to *bcoayevha* if the system shuts down?	Sorry, do you need to do *what*?
2	Do you need a *hadsfiywehfdsakhas* capacity?	Sorry, do we need a *what* capacity?
3	What is the point of the *nvjciuotakljejaiuy* in the design of the architecture?	Sorry, what is the point of the *what* in the design?
4	Can you tell us more about the *haslhfdhkao shojhliafs kjlhasd fjkllkjh afskljh*?	Sorry, can I tell you more about the *what*?

The strategy is:

*   begin your request for clarification with *sorry*. This immediately signals to the questioner that you have not understood something
*   substitute the part that you did not understand with *what*.

15.11 Learn strategies to help you understand and clarify questions (cont.)

You can use *what* to substitute any part of speech: a verb (example 1), and adjective (2), a noun (3), or even a complete phrases (4). By clearly identifying what it is that you did not understand you encourage the questioner not to repeat the whole question, but only that part of the question that you did not understand. This thus increases your chances of understanding.

If you have tried twice to get clarification and you have failed, then you need to admit your failure and say:

I am sorry, I still don't understand. Is there anyone else here who can help?

Sorry, maybe you could email me the question, and I will make sure I get the right answer to you.

On the other hand, if you manage the question, you can say:

Does that answer your question?

Is there anything you would like me to repeat?

To learn more about how to understand native English speakers, see Chapters 13 and 14 in the companion volume *Meetings, Negotiations and Socializing*. To improve your listening skills, read Chapter 15 in the companion volume *Telephone and Helpdesk Skills*.

15.12   If you are co-presenting, make sure you agree beforehand how to answer questions

Sometimes you may give a demo with a colleague, and you will both thus have an opportunity to answer questions. It is very important that you are consistent with your answers and that you don't show any signs of disagreement; you don't want to lose credibility in front of your audience.

The secret is to predict all the questions that you might be asked (see 15.1). Then you can agree together on how you will answer them.

If you are asked a question whose answer you have not prepared, then decide on the spot who is going to answer the question. You can say:

> *Shall I answer that question?*

> *I think you are best answering that question.*

15.13   Terminate the Q&A session if you feel you are wasting time

If you have questions during the demo but feel that enough time has been spent on them, show the next slide, or write something on the whiteboard. This signals that you are going to move on to the next topic and this should stop the audience from asking further questions.

Prepare for questions taking too much time by deciding in advance which slides can be skipped so that you can do the rest of the demo more quickly.

For tips on how to generate and deal with questions in demos and training sessions see Chapter 13.

16 PRACTISING YOUR PRESENTATION

16.1 Don't improvise. Practise exactly what you are going to say

If you have not written a script (see Chapter 4) and / or have not practised what you are going to say, you will probably fill your speech with redundancy. Imagine you have developed a system for helping people trade on financial markets. Your slide looks like this:

System for Financial Markets

1. Dealer side

2. Central Market

3. Customer Side

Compare these two versions of what you could say.

IMPROVISED	SCRIPTED / REHEARSED
As you have probably seen from the technical documentation, the system has been divided into three main parts. The first one is the dealer side and then there is the central market and in the end is the customer side.	As you can see, we have divided the system into three main parts: dealer side, central market and the customer side.

A. Wallwork, *Presentations, Demos, and Training Sessions,*
Guides to Professional English, DOI 10.1007/978-1-4939-0644-4_16,
© Springer Science+Business Media New York 2014

16.1 Don't improvise. Practise exactly what you are going to say (cont.)

The improvised version contains twice as many words as the rehearsed version. This means

- it will take twice as long for you to say

- half of what your audience hear will be redundant

You are thus wasting your time and theirs. You are also wasting an opportunity either to give them more content or to check that they have understood the information you have given them.

Instead, if you have prepared a script, your initial practice could simply be to read your script aloud so that you become familiar with what you want to say. Then, abandon your script completely, and just use notes.

As you practice, if any phrase or word does not come easily to you, try to modify it until what you want to say comes quickly and naturally.

Finally, try doing the presentation aloud without looking at your notes. Of course, if you forget what to say, then quickly look at your notes.

Even the best presenters make use of notes on the day of their presentation – it is standard practice and no one will think it is unprofessional if you occasionally look down to remember what you want to say.

16.2 Practice your position relative to the screen

Try to reproduce the real conditions of the room where you will be doing the presentation. So if you are practicing with colleagues don't stand right next to them, but at a distance.

Imagine the screen is behind you. Think about the best place to stand. If you stand in front of the screen, the beam will light you up and audience won't really be able to see you.

To avoid blocking your slides from the audience's view, stand to one side of the screen. Only move in front of it when it is strictly necessary to point to things on your slides.

Note that if you stand on left side of the screen, you will probably focus just on those members of the audience on the right hand side (and vice versa). So you need to keep swapping sides.

Make sure you make eye contact with everyone including those at the back. If you don't give certain sections of the audience regular eye contact, they will start to lose interest.

You can practice this by yourself at home. Stand at one end of the biggest room of your house. Imagine that the items of furniture (chairs, tables, desks, shelves, even windows) in various parts of the room are members of the audience. Practice talking to each item. Spend no more than two seconds on each item of furniture then move on to another item. In fact, if you focus on a single individual in the audience for more than two seconds, this individual will feel uneasy.

It also helps if you can project your slides onto a wall. This will help you become familiar with learning not to look at your slides, but at the audience. Of course, if you only have a minimal number of slides that you don't really need to look at (because they are so simple or easy for you to remember) then you will have less temptation to look behind you. In any case, you should be able to deliver your first 60 seconds without looking behind you at the screen, or at your laptop or at your notes.

16.3   Don't sit. Stand and move around

It is a good idea not to sit and talk into your laptop. When you are sitting your voice does not project as well.

You can also make better eye contact with people further back simply by leaving your laptop and moving around the room. This will also help you feel more relaxed. It is also an excellent way of gaining the audience's attention rather than the screen being their focus of attention. But make sure there are no wires in your path as you may trip over them.

If you move in a relaxed, but not repetitive, manner in front of the audience, it will give them the impression that you are at ease and comfortable in the presentation environment. And by implication your ease will make the audience think you are confident about your presentation itself.

If you seem relaxed, the audience will be too. They will thus be more receptive to your ideas.

Standing in a different position once every two or three minutes, will also help you to remember not to focus on just one section of the audience.

16.4   Use your hands

Do whatever comes most naturally to you with your hands and arms. Inexperienced presenters often begin by rigidly holding their arms to their side, or folding them across their chest. Such positions tend to make the audience feel that you are nervous or may be a bit hostile. So try and move your hands around as soon into the presentation as possible. A perfect point to do this is in your Outline, where you can use your right hand to touch the fingers of your left hand to indicate your three / four main points, e.g. by saying *first I want to, second.. third.*

Some people find it helps them overcome their nervousness by holding something in their hand, for instance the remote control, a pointer, or a pen. Try only to do this for a few minutes, as it stops you making full use of your hands.

Others find they are more relaxed with their hands in their pockets, but this may make the audience feel that the presenter is not very professional.

In any case, avoid things that may be distracting for the audience such as playing with your ring or scratching any part of your body.

Many good presenters use their hands to add extra emphasis to what they are saying. However, if for cultural reasons you feel that using your hands would be a sign of disrespect or lack for professionalism for the majority of the audience, then do what feels comfortable for you.

16.5 Have an expressive face and smile

If you show interest in your face then the audience will feel it and will themselves become more enthusiastic about what you are saying.

If you just have a blank expression, then you will not transmit any positive feelings to your audience.

You can practise smiling in front of the mirror. You can also video yourself practising your presentation. You can then see how often you smile, and if necessary then make an effort to smile more often in your daily life.

16.6 Vary the parts you practise

Given time constraints, people often manage only to practice part of their presentation at a single time. The result may be that you only practice the first half of your presentation. So it is a good idea to occasionally begin in the middle, or begin with the conclusions – don't just focus on the technical part. Also, don't forget to practice answering questions – imagine the question, and then answer it in various ways (including imagining that you didn't understand the question).

In any case, practice the opening and the ending again and again and again. These are the two parts of the presentation where you should not improvise, and where it helps considerably if you know exactly what you are going to say. First and last impressions are the ones that remain with the audience.

16.7 Use shorter and shorter phrases

As you practice try and make your phrases shorter and shorter. Short phrases give you time to pause quickly and to breathe between one phrase and the next – this will slow your speed down if you are nervous.

16.8 Prepare for forgetting what you want to say

A frequent problem is forgetting a specific word or phrase that you need to say.

There are three good solutions for this, you can:

- look at your notes (either on paper or on your mobile phone / tablet - see 9.18)

- drink some water, or take out a handkerchief to wipe your nose, and use this time to remind yourself

- say *I am sorry I can't think of the word. In any case...* And then you simply proceed with the next point

16.9 Opt to do presentations in low risk situations and put yourself at the center of attention in social situations

Begin your presentation career by presenting in low risk situations, for example internal presentations in front of colleagues.

In addition, if you are the kind of person who prefers not to be the center of attention, then try to overcome this fear by gradually building up your confidence. You can do this by:

- participating more actively in work meetings, e.g. voicing your opinion rather than remaining silent

- taking part in activities outside work such as team sports, acting clubs, speakers clubs

- beginning conversations with strangers on buses, trains and planes

16.10 Learn how to be self critical: practice with colleagues

Learning to be able to evaluate your own demo and your presentation skills is key to getting your message across to your audience. If you ask your colleagues *How did I do?* or *What do you think?* they will probably just give you some vague encouraging comment. Instead it helps to have a check list.

Also, bear in mind that the things you find ineffective in your colleagues' presentations, may be exactly the same kind of mistakes you make, so you can certainly learn from other people's errors.

Below are some points that you may find useful to include in a check list. For each point, you could ask your colleagues to rate you: 1 poor, 2 satisfactory, 3 good

Assessment Sheet	
STRUCTURE	VOICE / DELIVERY
• Strong beginning – topic introduced clearly	• Right speed – did not begin in a rush
• Clear conclusions and strong ending	• Overall topic previewed
SLIDES	• Clear transitions and links between points
• Clear text	• Clear and loud voice
• Simple diagrams	• Short clear phrases, individual words articulated clearly
• Not too much detail	• No annoying noises (*er, erm, um*)
• No distracting colors, fonts, animations	• Good pronunciation
BODY LANGUAGE	• Enthusiastic and friendly
• Eyes on audience not on screen	• Sounded credible
• Moved around	AUDIENCE INVOLVEMENT
• Used hands appropriately	• Attention of audience immediately gained
	• Topic clearly related to audience
	• Audience personally involved in some way
	• Variety to maintain attention

16.10 Learn how to be self critical: practice with colleagues (cont.)

In addition you should ask you colleagues whether you have any annoying tics. A tic is an involuntary action. Audiences can be distracted by presenters who:

- say *er, erm* and *uh* between one word and another

- say *OK?* very frequently

- make any strange bodily movements, for example if when sitting down you bounce your leg up and down

- touch their face very frequently

- constantly clear their throat

The problem is that you probably are not aware of whether you do any of the above.

16.11 Think about what slides you could delete

Frequent participants at presentations complain of the phenomenon of 'death by powerpoint' – slide after slide after slide simply sends audiences to sleep. What has happened is that the only visual aid used nowadays is a slide with a text, graph, table or picture on it. Thus it may be that to clarify a difficult point or to be more persuasive, you actually have to reduce the number of slides and use other methods instead.

When you have practised several times, think about what slides you could delete in order to streamline your presentation.

A useful feature of PowerPoint is that you can print up to nine slides on a page – this is called 'print as handout'. When you see all your slides together like this, it gives you a clearer picture of the amount of text you have used throughout your presentation.

Look at each slide and ask yourself if the text is crucial. If it is not crucial, cut it.

If it is crucial then ask yourself – can I express it in a more succinct way? Could I use a picture rather than text? Do I really need a slide to express this point or could I just say it verbally?

Next time you watch someone doing a presentation, decide if their slides were:

a)   specifically designed to help the audience understand the topic

b)   or simply prompts for the presenter so that he / she wouldn't forget what to say next

The main task of your slides should to be fulfill point a), and at the same time fulfill point b).

Try to reduce any overlap between what you say and what your slides 'say'. The slides do not need to contain everything that you will say. You just need a slide for your most important points. And it is your job to draw the audience's attention to why the information on your slide is important.

A slide should only come alive when you actually start commenting on it. Slides should support the talk, not reproduce the talk itself in a written / graphical format.

16.12   Email your presentation to your boss and colleagues

Make sure that your boss and colleagues see your presentation or demo before you actually deliver it in front of your client. The solution is to email him / her the presentation beforehand so that they will then know what to expect. What you don't want is your boss putting their hands in their hair in despair or your colleagues looking confused. However, your boss is likely to be busy and probably won't read the presentation. Instead just say 'Here is my presentation. Could you just check slide 20 as I am not sure how the audience will react'.

This is a good strategy if you are not sure, for example, whether:

- a graph or figure will be clear to the audience

- you have written too much text (or not enough)

- a humorous slide might be inappropriate

- you have covered everything that needs to be covered

16.13   Learn relaxation techniques

As you know from taking exams, being slightly nervous actually helps you to perform better. If you are too relaxed you become over-confident. Don't worry about your nerves, they will soon disappear a couple minutes into the presentation.

Make sure you sleep well the night before. Don't stay up all night rewriting your slides. You should arrive at the presentation feeling fresh, not tired. If you feel stiff and rigid at the beginning of a presentation you may need to learn some relaxation techniques.

Do some physical exercises before you begin:

* breathe in deeply

* relax / warm you neck and shoulder muscles

* exercise your jaw

Another useful exercise is to make your body appear big in the minutes before the demo or presentation. Frightened animals often scare off predators by making themselves look bigger than they really are. Competitive runners often raise and stretch their arms high in the air in kind of victory salute as they cross the finishing line. Again, they are making their body appear bigger. Research has shown that if you practice extending your body before, for example, an interview or presentation, then you will perform better. You feel big so you put on a big performance. So don't sit down in a corner before your presentation, but stand up and adopt a 'big' pose.

16.14   Check out the room where your presentation will be

It is a good idea to familiarize yourself with the room where you are going to be doing the presentation. Try to imagine yourself in the room doing your presentation. Then think about / find out about:

- how loud you will have to speak given the size of the room and how far you are from the audience

- whether you will need a microphone

- where you will position yourself so that the audience can always see you and so that you don't trip over any wires

- how the remote control works e.g. how you can blank the screen without turning the projector off (the button is generally called 'blank', 'no show', or 'mute'); and how effective the laser pointer is

- where chalk and pens are available for the blackboard / whiteboard

- whether bottles / cups of water will be provided

16.15   Prepare for the software or the equipment breaking down

Go early to the room where the demo / presentation will be held. Check that all the equipment works correctly as much time in advance of your presentation as possible. This is important as there are different software versions and sometimes incompatibilities between Macs and PCs (particularly regarding animations).

Some of the most successful presentations are done with no slides. If you have a print out of your slides and your computer breaks down completely then you can continue without the slides, and if necessary draw graphs on a whiteboard.

In any case, it is a very good idea to practice for such a breakdown, i.e. to give your presentation without any slides. It will teach you two things:

- it is possible to do a presentation with no slides

- it will show you which of your slides are probably redundant.

16.16   Organize your time

Presentations rarely go to plan. So allow for:

- the previous demo or presentation going over their allocated time, meaning that you have less time to do your own demo

- people arriving late

Prepare for this by:

- knowing exactly how much time you need for each part of your presentation

- having your most important points near the beginning of the presentation, never just in the second half

- thinking in advance what slides you could cut, particularly those in the latter part of the presentation

- planning how to reduce the amount you say for particular slides

- using options in your presentation software that allow you to skip slides

You cannot calculate the length of your presentation from the number of your slides. For example, if you are doing a 10 minute presentation you may spend two minutes on the title slide as you introduce yourself and the topic. Then the next slides may entail you giving quite long explanations of the background. Also, there should be parts in your presentation in which you do not need slides. This means that there is not an average time for each slide. So you need to time the whole presentation to see how long it takes, and then decide where you could make cuts if on the day you don't have your full allocated time.

If you do run out of time don't suddenly say: *I will have stop here*. Instead, briefly make a conclusion.

17 IMPROVING YOUR SKILLS AS A PRESENTER AND TRAINER

The more presentations, demos and training sessions you do, the more proficient you will become. This chapter is intended for those of you who already have some experience but wish to improve your skills.

17.1 Learn how to gain and keep your audience's attention

Below are some good ways of attracting and holding your audience's attention:

1. have a clear idea who your audience are, don't assume that they are naturally going to be interested in your topic

2. have an agenda and a clear structure with clear transitions so that the audience know where you are going

3. make it easy for the audience to follow you and your slides

4. help the audience to understand why you are showing them a particular slide

5. involve your audience and give them lots of examples

6. make frequent eye contact

7. avoid too much text on your slides

8. use simple graphs and tables

9. make your text and visuals big enough for everyone in the audience to see clearly

10. avoid entering into too much detail (i.e. just select those things that the audience really need to know about the topic)

11. avoid spending more than a couple of minutes on one specific detail

12. have a variety of types of slides (not just all bullets, or all text, or all photos)

13. speak reasonably slowly and move from slide to slide at a speed that the audience will feel comfortable with

14. vary your tone of voice

15. occasionally stop talking for five seconds or more

A. Wallwork, *Presentations, Demos, and Training Sessions,*
Guides to Professional English, DOI 10.1007/978-1-4939-0644-4_17,
© Springer Science+Business Media New York 2014

17.2   Don't make your agenda look unfeasible

When you are outlining your agenda ensure that audience understand that you are not going to be speaking in detail about every point on the agenda (see 5.6). For example, even if you have eight points on your agenda, you can say something like

Don't worry, we are only going to spend time on the second and third points, the others I will be going through quite quickly as I think you will already be familiar with them.

17.3   Maintain eye contact with the audience

Don't hide behind your laptop – maintain as much eye contact as possible. If you don't make eye contact with all your audience throughout the duration of your presentation, they will quickly start thinking about other things. Check the audience's body language to see whether they are understanding and whether you are going at a reasonable speed.

You can only maintain eye contact with the audience if:

- you know exactly what you are saying – if you are not sure what you are going to say next, you will probably start looking up to the ceiling or down to the floor

- your slides are simple – if they are complex you will be very tempted to turn your back to the audience to remember the information on the slide

17.4   Exploit moments of high audience attention

Audiences tend to remember things that are said at the beginning and end of a presentation, because their attention is generally high at these points.

They also remember things that they hear more than once.

And finally they remember curious facts, i.e. things that stand out.

Ideally you need to state your key points both at the beginning and ending. In the middle go through each key point more in detail. If possible, include an unexpected / counterintuitive / interesting fact for each key point. Try juxtaposing data with quotations, and serious issues with a humorous anecdote.

17.5 Don't tell the audience everything you know, only what they need to know

Try to avoid the temptation to give the audience the full Wikipedia explanation. When you've written out your speech (Chapter 4) for the first time, revise it, and if possible reduce the amount you have written.

Don't state the obvious. If your slide is entitled 'Roadmap' and shows a list of milestones that have already been done and things that will be done in the future. Don't say:

> *This is a roadmap of the XXX project. We have already achieved the first two steps and we now have a beta version with full functionality. The plan is to have final version, Q1, next year.*

Instead just point to the relevant parts of the slide and say:

> *This is what we've done already, and the plan is to have the final version early next year.*

17.6 Don't refer to everything on your slide

Be as economical as you can with the parts of your presentation / demo that:

- are not giving key information
- are clear from the slide and need no further explanation

Let's imagine you are referring to some enhancements you have made to an existing product. You have a slide entitled General Improvements (see below), which acts as a mini agenda for the next series of slides in which you will explain what improvements you have made on the old product. You only need to mention the first item (increased speed).

GENERAL IMPROVEMENTS

- Increased speed
- Enhanced features
- More levels of security
- Lower maintenance costs

You could say:

First I want to take you through what we have done to increase the speed, and after we'll have a look at these three improvements.

You then simply point to the three other bullets.

In summary, if you have a series of five or six bullet points, just comment on one or two of them – the audience does not need or want an explanation of all of them.

17.7   Avoid details / exceptions etc

One problem you may have is thinking that you have to include all the details because this is what your boss or the audience expect of you. You may think that by leaving them out, you will be considered to be unprofessional or worse not to know about your job. If you give all the details, you will force your audience 1) to hear extremely complex explanations that cover all possible cases, and 2) to look at extremely complex tables and graphs.

Don't worry about leaving out the details. Just introduce what you say with a qualification:

> *This is an extremely simplified view of the situation, but it is enough to illustrate that ...*

> *In reality this table should also include other factors, but for the sake of simplicity I have just chosen these two key points:*

> *Broadly speaking, I think we can say that ...*

You can then tell the audience that the details are provided in the support documentation (in the handout, on your company's website etc).

17.8   Don't spend too long on one slide

Our attention span is affected by how long we look at something that does not change. Research has shown that we can only look at something static for 30 seconds and then we start thinking about something else. So if possible, reduce the amount of time you spend showing the same slide. For example, you could show the slide, explain what you need to explain with the aid of the slide, and then blank it and carry on talking.

17.9 Learn techniques for regaining the audience's attention

When you are doing your presentation you may be competing for the audience's attention with one or more of the following:

- their mobile phone or laptop – they may be text messaging or emailing

- the person sitting next to them who may want to chat

- things happening outside the window

- their hunger (particularly in the late morning or late afternoon)

- their boredom – yours may not be the first presentation they have seen that day

Also, no matter how exciting you think your product or service is, audiences can only concentrate for a certain amount of time (probably about 15 minutes). You can regain their attention by:

- blanking the screen (on PowerPoint you can do this using the B key)

- using the whiteboard – inevitably the audience will want to know what you are going to write. Make sure you write large enough for all the audience to see – this generally means writing very little or only drawing simple diagrams. Move to the side of the whiteboard so that the audience can see what you are writing / have written

- asking the audience a rhetorical question. Try and predict what kind of questions the audience might be asking themselves at this point of your presentation. Pause. Ask the question. Pause again. Then answer it

- giving the audience a statistic. People are fascinated by numbers and they help the audience see the dimension of the situation

- saying *here's something you might be interested in seeing* or *I've brought along something to show you*..and produce an object from your pocket, bag etc. Your audience will be immediately curious to see what the object is. Again it has to be big enough for everyone to see, or you have to have lots of them to distribute among the audience – but be careful as they may turn into an even bigger distraction! Objects can also be a good substitute for explanations

- showing an unusual slide – this could simply be a slide that breaks with the normal pattern of your other slides. It could be an interesting photo, a clear and effective diagram, or contain a number, a short quotation, or a question

17.10 Make sure what you say contains new information

If you have a slide that you have decided is worth including because it contains important information, you must decide whether the audience will already be familiar with the concept. If they are familiar, don't spend too long on the slide.

For example, if you are describing what some software or some instrument will do, focus on what makes it different from other software and instruments. Don't go into detail about what it does that other similar products and tools do. If you do, the audience will soon lose interest.

17.11 Present apparently old information in a new way

One major problem is that a few minutes into your presentation the audience may decide they have either already heard it all before or that they already know it. If this really is the case, then this means that you haven't found out about their needs.

However, generally the problem is that the audience doesn't really listen to what you are saying and they think they have heard it before whereas in reality they haven't. Also people have a tendency to predict what you are going to say next and are thus inclined to hear what they predicted you would say rather than what you actually said.

This means that you constantly have to underline for them why what you are saying is really important for them. You can do this by apparently reading their minds and saying out loud what they are thinking:

Now I know you think you have heard this all before but ...

I know that you already have something that you probably think is similar but ...

You are probably thinking that I am just going to... but in fact what I am offering is something radically different which will really transform the way you do ...

You can also present standard information in an unexpected or novel way, or by juxtaposing apparently unconnected statistics to show how they are in fact related and thus prove your point.

17.12 Underline relevance and value

A very frequent mistake is for you to assume that your audience will automatically connect what you are saying to their own situation. Your job is first to make sure that there really are connections, and secondly to help the audience see these connections. Show them how your statistics specifically apply to them. For example, instead of saying:

20% of people in this new company will lose their job if ...

You could say:

20% of those of you will lose your job if ... that means at least two people in this room.

But remember that not everyone is convinced by statistics. Always try to refer to the personal experience and feelings of the people you are addressing – give them a 'soft' proof of what you are saying. For example, your software is designed to make people's tasks more efficient and to make their lives and the lives of others easier. So when explaining to IT people from banks how your software works, think in terms of how it will improve their relationship with the traders, or how it will save them time. These are the things that are really important for them on an everyday basis.

Whenever you can, make comparisons with what you know the audience experience in their working lives. If you really want to convince your audience, try to make the benefits of what you are saying as specific as you can. Any vague benefits you mention may have the opposite effect and make the audience think that you are just inventing them.

17.13 Explain or paraphrase words that may be unfamiliar to the audience

Make sure the audience understand key words – explain / show what they mean, as a multilingual audience may know the concept but not the word in English.

If you use a non-technical word which you think the audience may not know, say it and then paraphrase it. Example: *These devices are tiny, they are very small.*

17.14   Repeat key words and concepts frequently

Think about how you learn English – can you always remember (and apply) a rule the first time you hear it? It is very important in a presentation to recap what has gone before. This may seem very boring to you, because you know your presentation so well. But for the audience this may be the first time they hear something.

So, don't be afraid to make the same point twice, but try and express it in a different way. For example, if you are telling people how to make money, you can re-explain the concept in terms of how not to lose money.

This will enable people to:

- follow what you are saying better,
- catch concepts the second time if they missed them the first time
- remember afterwards what you have said

17.15   Use breaks

Breaks give you time to relax a little, to collect your thoughts, and to make any adjustments to what you are going to show participants next.

Breaks also allow the audience to relax. There is only so much information that the audience can assimilate at one time.

It is useful to appoint one of the participants to round up the others when the allocated time for the break is over.

17.16 Enjoy yourself: Sound like you are talking to a colleague

Don't confuse being professional with being detached from the audience and adopting a particular tone and voice. Talk to the audience as if you were talking to a colleague – you don't need to go into 'presentation mode'.

For example, listening to how something works and what it can do is often really tedious for the audience, particularly if you cannot give them an opportunity to try the thing out. To avoid this problem, you can use a narrative style (i.e. as if you were telling a story) to describe even a piece of software or equipment. Instead of saying:

This part of the application allows you to do X. Then as you can see in this slide, this part allows you to do Y. And with this part here you can do Z. etc etc etc

You can talk about the story and rationale behind each of these parts.

A lot of users told us that they spent a lot of time doing X but had no quick way to do it. Well now they have, as you can see here. Possibly the most difficult part of designing the application was thinking of a way to do Y, and I hope you like the solution we have come up with ... Then with Z we found we couldn't do ... so then we tried ... but that didn't seem to work ... finally we discovered that ... then we found ourselves up against another problem ... but meanwhile the GUI group had come up with a fantastic solution, which is the one you can see here.

Clearly your presentation would take several hours if you used this style all the time. But after the middle of the presentation when your audience's level of attention is likely to be at its lowest, this technique will certainly regain their interest.

Below is another example.

Imagine that you work for an IT company. You are explaining to your client how your software, Eagle Eye, will help them increase their business vision. Compare these two ways.

NORMAL	DYNAMIC
The size of a bird's eye and a human's eye take up about fifty per cent and five per cent of their heads, respectively. In birds of prey, such as eagles, vision is of crucial importance. So too is vision in the business world. Our software application, which we have denominated Eagle Eye will provide you with a better vision of your competition and gaps in the market.	*A bird's eye is huge. It takes up about fifty per cent of its head. Half its head. That's ten times more space than a human's eye takes up. To be comparable to the eyes of a bird of prey, such as an eagle, our eyes would have to be the size of a tennis ball. With Eagle Eye, you will get ten times better vision of your competitors than you are getting now. Gaps in the market will suddenly become ten times clearer.*

17.16 Enjoy yourself: Sound like you are talking to a colleague (cont.)

Note how in the revised version, the speaker gives the same information twice – *fifty per cent* and *half*. This is useful because it is very difficult to distinguish between the sound of *fifteen* and *fifty* (likewise between 13 and 30, 14 and 50 etc). By using the analogy of a tennis ball, the audience gets a much clear idea of the proportions. This then makes it clear how their business opportunities will be ten times greater with Eagle Eyes.

Now compare two ways to explain how Eagle Eye can be integrated into the client's existing system.

ALIENATING AND REDUNDANT	ENGAGING
So basically what I am saying is that the architecture of the system and the way it has been implemented should lead to a reduced quantity of work with regard to the integration, thus giving you the possibility to have a fully featured system up and running within the context of a very limited time frame.	*What we've tried to do is to limit the amount of integrating that you'll have to do. The upside is that you'll be able to get a fully featured system running in a very short time.*

17.17 Inject some humor

This doesn't mean telling jokes, but simply:

- making humorous analogies (e.g. comparing your product or their job with something unexpected)

- explaining something in an unexpected way

- making friendly references to problems that the audience frequently experience in their everyday working lives

- recounting amusing events that surrounded the development process of the product you are explaining

- telling an anecdote that in some way relates to the audience

Using humor has several advantages. It:

- involves the audience, creates a bond with them, and makes them feel good

- improves the audience's impression of you

- creates variety in your presentation and stops it from becoming too somber

Some nationalities and some people like to be entertained during a presentation. However, telling jokes may be dangerous as the joke may:

- not be understood

- be offensive or inappropriate for the culture of your audience

- be completely irrelevant to the topic of the presentation

But being entertaining doesn't always mean making people laugh. It means

- occasionally providing standard information in a novel or unusual way

- using examples that your audience can easily relate to

- finding interesting and surprising statistics

- using very simple but unusual graphs and pictures that underline important points in a new way

In any case you may decide to provide a few humorous slides or anecdotes. You can then try one and see what reaction you get from your audience. If it works well you can use the others. If not, skip them.

17.18   Don't talk for more than a few minutes at a time

Don't talk for more than about 90–120 seconds – ask the audience questions, get them to ask you questions (see Chapter 13). This will make the demo much more dynamic and interesting for both parties.

Use enthusiastic phrases such as:

We really value your help with this.

We are very excited about this ...

This is a feature that I personally think is especially useful

Of all the features we are planning to offer, this is the one that clients get most excited about ...

17.19   Give frequent examples

Imagine what questions the audience might ask – they might say ' can you give me an example of this?' or 'why did you decide to do that?'

Examples enable them to visualize the abstract concepts that you have given them.

17.20   Constantly remind the audience of the big picture

Make sure the audience is always aware of the big picture, by:

- Referring back to the agenda to show where you are
- Referring back to previous slides either verbally or by reshowing them the same slide.
- Reminding them why you are telling them something.
- Giving them mini summaries (remember you are very familiar with what you are talking about, but they need reminding)
- Warning them about what's coming next.

17.21 Write new slides just before the presentation begins or during a break

A clever technique for gaining audience attention is to create ad hoc slides using comments that you have heard as you have been wandering around the room before the presentation or based on questions that people have asked you during the presentation. Clearly, you can only do this if you have a break in the presentation or if you give the audience some task to do.

Alternatively you can write their comments on a whiteboard. Then later as you go through the presentation, you can refer to them. In this way the audience understand how what you are saying relates directly to their needs.

17.22 Improve your slides after the presentation

When you do your presentation live in front of a real audience it sometimes reveals faults that did not appear while you were practising. Look at your slides with a critical eye and ask yourself:

- why was this slide necessary? if I cut it, what would change?
- did this slide really support the objective of my presentation?
- why did I include this info? was it relevant / interesting / clear? what impact did it have?
- could I have expressed this info in a clearer or more pertinent way?
- was this series of slides in the best order? was there anything missing in the series?
- were these slides too similar to each other? did they really gain the audience's attention?

After your presentation, write down the questions you were asked, so that the next time you do the same presentation you will have the answers ready.

18 USEFUL PHRASES: ALL TYPES OF PRESENTATIONS

The following chapters contain lists of useful phrases that you can you use in your own presentations, demos and training sessions.

Chapter 18: phrases that are useful for all kinds of presentations

Chapter 19: informal technical demos

Chapter 20: formal business presentations, possibly with large audiences

For each stage of a presentation, you will find several alternative phrases. Choose the ones that you find easiest to:

- remember

- pronounce

It is also worth becoming familiar with the other phrases. They will help you improve your understanding when you are an attendee rather than a presenter.

18.1 Preliminaries and introductions

Reminders before starting

Before we start, I'd like to draw your attention to …

One small reminder – please switch off your mobile phones.

Could you please keep your questions until the Q&A session.

Referring to documents needed to follow the call / presentation

If you are listening on the phone, the slides we are going to be showing are available to download from *website address*.

Those of you on our email list should have already received a copy of the slides. If you haven't, and would like to be placed on our list, please let us know.

You will find all of the information I will present, along with further supporting detail, on our website.

I am accompanied today by my colleagues and all of our biographies can be found on pages 3 and 4 of the report.

A. Wallwork, *Presentations, Demos, and Training Sessions,*
Guides to Professional English, DOI 10.1007/978-1-4939-0644-4_18,
© Springer Science+Business Media New York 2014

Getting started

Okay, we'll kick off. Good morning everybody and thanks very much for coming.

Before walking you through *topic*, let me describe the ...

I'll start by summarizing ...

I'd like to take you through the highlights of our ...

Firstly, I am going to give you some background in the areas of ...

Introducing yourself

My official job title is ...

My role within the company is to ...

I deal with matters regarding ...

I am responsible for ...

It is my job to ...

I report directly to ...

I began working for *company* in 2015.

I have been with *company* since 2015.

I am based in *place*.

18.2 Outlining agenda

Giving agenda / outline

I will begin by giving you an overview of ... Then I will move on to ... After that I will deal with ... And I will conclude with ...

First, I'd like to do x ... then I'm going to do ...

First, I'll be looking at x. ... Then we'll be looking at y ... After that, we'll focus on z. ... And finally we'll have a look at ...

Finally, I'm going to take you through z.

So, let's begin by looking at x.

So this is what I am going to talk about ... and the main focus will be on ... and what I think about ...

What I hope you will find interesting is ...

I'm NOT going to cover ... I'm just going to ...

The presentation is about 45 minutes and we've got about another half hour for Q&A after that. Then there will be a lunch break for one hour.

This afternoon's session will start at two o'clock – please be prompt.

Negotiating with your participants on what you will cover

Are there any points that you think we don't need to cover?

Is there anything you would like me to focus more on?

Are there any other any points that you would like me to include?

Giving your agenda (more dynamic)

This is what I'm planning to cover.

I've chosen to focus on X because I think it has massive implications for … and it is an area that has been really neglected …

I think we have found a …

  radically new solution for …

  truly innovative approach to …

  novel way to …

Why is X is so important? Well, in this presentation I am going to give you three good reasons …

What do we know about Y? Well, actually a lot more / less than you might think. Today I hope to prove to you that …

18.3   Moving from slide to slide and topic to topic

Making transitions

Looking ahead to …

As noted earlier, …

Before moving on to the other changes, let me summarize.

I'd now like to turn to …

Turning to …

Let's turn our attention to …

In conclusion …

In summary …

So just to recap …

Let me now move onto the question of ...

Next I would like to examine ...

This leads me to my next point: ...

Signaling that the topic is about to change

In a few minutes I am going to tell you about X and Y, which I hope should explain why we ...

Before I give you some more detailed statistics and my overall conclusions, I am just going to show you how our results can be generalized to a wider scenario.

But first I want to talk to you about ...

I'd like to now move from x-related matters, and cover several other changes.

In order to help you appreciate the high performance of ..., I will now ...

Referring backwards and forwards

I'm not going to cover this aspect now, I'm just going to ...

I'll go into a bit of detail for each concept.

I'll explain this in a moment.

I'll talk about that later.

We'll come back to this point later.

I'll say more about this later.

You may recall from an earlier slide that ...

As I said before ...

Remember I said that ...

The concept I mentioned earlier ...

As I mentioned a moment ago ...

To return to my earlier point ...

If we go back to this slide ...

Referring to previous topic to introduce next topic

Before moving on to z, I'd just like to reiterate what I said about y.

OK, so that's all I wanted to say about x and y. Now let's look at z.

Having considered x, let's go on and look at y.

Not only have we experienced success with x, but also with y.

We've focused on x, equally important is y.

You remember that I said x was used for y, well now we're going to see how it can be used for z.

Showing where you are in the original agenda

OK so this is where we are ...

This is what we've looked at so far.

So, we're now on page 10 of the handout.

Making mini summaries as a means of transition

Over the last few slides I've talked about ... , and four reasons why we think there are very interesting opportunities in ...

The next few slides cover the four reasons why we think we ...

So, I've covered the opportunities and why we think ...

Now let's talk about ...

That completes what I wanted to say about ...

18.4   Emphasizing, qualifying and explaining

Emphasizing a point

I must emphasize that ...

What I want to highlight is ...

At this point I would like to stress that ...

What I would really like you to focus on here is ...

These are the main points to remember:

The main argument in favor of / against this is ...

This is a particularly important point.

This is worth remembering because ...

You may not be aware of this but ...

Communicating value and benefits

So, the key benefit is:

One of the main advantages is ...

What this means is that ...

We are sure that this will lead to increased ...

What I would like you to notice here is …

What I like about this is …

The great thing about this is …

Qualifying what you are saying

Broadly speaking, we can say that …

In most cases / In general this is true.

In very general terms …

With certain exceptions, this can be seen as …

For the most part, people are inclined to think that …

Qualifying what you have just said

Having said that …

Nevertheless, despite this …

But in reality …

Actually …

In fact …

Giving explanations

As a result of …

Due to the fact that …

Thanks to …

This problem goes back to …

The thing is that …

On the grounds that …

Giving examples

Let's say I have … and I just want to …

Imagine that you …

You'll see that this is very similar to …

I've got an example of this here …

I've brought an example of this with me.

There are many ways to do this, for example / for instance you can …

There are several examples of this, such as …

18.5   Describing slides and diagrams

Introducing diagram

This is a detail from the previous figure …

This should give you a clearer picture of …

This diagram illustrates …

Here you can see …

I have included this chart because …

I am sorry that this diagram is so small, but I really only need you to notice x, y and z.

You can see this figure more clearly if you turn to page 3 of your handout.

Explaining what you have done to simplify a diagram

This is an extremely simplified view of the situation, but it is enough to illustrate that …

For ease of presentation, I have only included essential information.

For the sake of simplicity, I have reduced all numbers to whole numbers.

In reality this table should also include other factors, but for the sake of simplicity I have just chosen these two key points:

This is obviously not an exact picture of the real situation, but it should give you an idea of …

I have left a lot of detail out, but in any case this should help you to …

If you are interested you can see the full picture in the handout.

Indicating what part of the diagram you want them to focus on

Basically what I want to highlight is …

I really just want you to focus on …

You can ignore / Don't worry about this part here.

This diagram is rather complex, but the only thing I want you to notice is …

Highlighting details

As shown on the top line …

The table shows …

The table at the bottom of the slide shows …

Here you can see …

216

Notice that it has …

As you can see …

On the left is … On the right-hand side …

Here in the middle … at the top …

Down in this section …

Over here is a …

The upper / lower section …

Explaining the lines, curves, arrows

On the x axis is … On the y axis we have …

I chose these values for the axes because …

In this diagram, double circles mean that … whereas black circles mean …
dashed lines mean … continuous lines mean …

Time is represented by a dotted line.

Dashed lines correspond to … whereas zig-zag lines mean …

The thin dashed grey line indicates that …

These dotted curves are supposed to represent …

The solid curve is …

These horizontal arrows indicate …

There is a slight / gradual / sharp decrease in …

The curve rises rapidly then reaches a peak and then forms a plateau.

As you can see this wavy curve has a series of peaks and troughs.

18.6   Dealing with problems

Equipment that doesn't work

I think the bulb must have gone on the projector. Could someone please bring me a replacement? In the meantime let me write on the whiteboard what I wanted to say about …

The microphone / mike doesn't seem to be working. Can everyone hear me at the back?

I don't know what has happened to my laptop but the program seems to have crashed. Please bear with me while I reboot.

Problems with the handouts

Sorry but the last two pages of the handout seem to be missing. If you email me, I will send them to you as an attachment.

Sorry, but the pages seem to have been stapled in the wrong order.

Sorry, I don't know what happened to the quality of some of the photocopies.

In the table on page 3, the last number in the second column should be 1000 not 100.

Apologizing

You know what, there's a mistake here, it should be …

Sorry this figure should be 100 not 1000.

Sorry when I said x I meant y.

I'm really sorry about that. I thought I had switched my mobile off.

Excuse me a second, I just need to get some water.

Sorry about that.

Sorry, what was I saying? Where were we up to?

18.7 Asking and answering questions

Referring to your level of English just before Q&A session

If you ask any questions I would be grateful if you could ask them slowly and clearly, as …

my English is a bit rusty.

many attendees here today are not native speakers of English.

Soliciting questions from the audience during the presentation / Beginning a Q&A session

We are going to have a break in about 10 minutes, but before that I would like to hear your questions.

Based on what we've looked at so far, what are your thoughts?

We've got time for a couple of questions before we move on.

What questions do you have at this point?

Does anyone have any questions on this?

I'd be really interested in hearing your questions on this.

One question I am often asked is …

Something you may have been wondering is …

Handling the session

OK, could we start with the question from the guy at the back. Yeah, you.

Sorry, first could we just hear from this woman / man at the front.

Do you mind just repeating the question again because I don't think the people at the back heard you.

I think we have time for just one more question.

OK, I am afraid our time is up, but if anyone is interested in asking more questions I will be here for another half an hour or so.

Questions that the audience may ask you

Could you go back to the last slide?

Could you give us more details about …?

Could you go over the diagram again?

Where can I get more information about …?

Can I just pick you up on something you said earlier?

I'm not sure I understand your point about .. Could you clarify it for me.

Can you give me an example of that?

Interpreting the questions

If I'm not wrong, I think what you are asking is …

Can I just be sure that I understand. You are asking me if …

So what you are saying is …

So your question is …

As far as I understand it, you are asking …

What to say when you don't understand a question from the audience

Sorry, could you repeat the question more slowly please?

Sorry, could you speak up please?

Sorry, I didn't hear the first / last part of your question.

Sorry, I still don't understand – would you mind asking me the question again in the break?

Sorry, but to explain that question would take rather too long, however you can find the answer in the product manual.

Avoiding difficult questions

I can't give you an exact answer on that I am afraid.

That's a very interesting question and my answer is that I simply don't know! But I will find out for and get back to you.

That's a good question and I wish I had an answer on the tip of my tongue, but I am afraid I don't.

You know, I've never been asked that question before and to be honest I really wouldn't know how to answer it.

I would not like to comment on that.

I am sorry but I am not in a position to comment on that.

Asking for time, or deferring

I think it would be best if my colleague answered that question for you.

Is there anyone here in the audience who can answer that?

Can I get back to you on that one?

Could we talk about that over a drink.

You've raised a really important point, so important that I think I would rather have a bit of time to think about the best answer. So if you give me your email address at the end, I'll get back to you.

Can I think about that and I'll give you an answer after the break.

Commenting on audience questions

I know exactly what you mean but the thing is ...

I take your point but in my experience I have found that ...

You're quite right and it is something that I am actually working on now.

Going back to the presentation after taking questions mid presentation

OK, would you mind if I moved on now, because I've still got a couple of things I wanted to say.

18.8 Ending the presentation / demo

Announcing that you are going to finish

OK, we're very close to the end now, but there are just a couple of important things that I still want to tell you.

It looks as if we are running out of time. Would it be OK if I continued for another 10 minutes? If any of you have to leave straight away, I quite understand. I am really sorry about this. But in any case, you can find the conclusions in the handout.

Well that brings me to the end of the presentation. So, just to recap:

Telling audience where they can find further information

I am afraid that I don't have time to go into this in any further detail. But you can find more information about it on this website.

If you would like more information on this, then please feel free to email me. My address is on the back page of the handout.

I will put a copy of the demo on our website.

Thanking the audience

Thanks very much for coming.

Thank you for your attention.

19 USEFUL PHRASES: INFORMAL TECHNICAL DEMO

19.1 Before the demo (face-to-face)

While you are setting up and someone asks you a question

Sorry, do you mind, but I am just getting the demo ready.

Actually, I've just got a couple of last minute things to do.

Sorry, I just need to go to the restroom.

Getting to know the audience as they arrive

So which department are you in?

How long have you been working in ...?

What do you already know about ...

Do you know how many people are coming today?

Are you all from the same department?

What are you hoping to learn during today's session?

I think we are all here now. Before we start, I'd just like to find out a bit about those of you who I haven't met yet. So let's start with you, can you just tell me what your role is here?

Could everyone just introduce themselves? Let's start with you.

Checking that everyone is present

Do you think this is everybody, or should we wait a few more minutes? Did word get round that the time of the demo had changed?

19.2 Before the demo (audio and video conference)

Presenter: technical questions and comments

Audio call At the moment I can see two people connected.

Video call Can you all see me OK? Is the audio loud enough?

Video call Do you think you could all move a bit closer to the camera.

A. Wallwork, *Presentations, Demos, and Training Sessions,*
Guides to Professional English, DOI 10.1007/978-1-4939-0644-4_19,
© Springer Science+Business Media New York 2014

Can you see the screen OK?

Is everyone picking up all right?

Is that any better?

Are you on speaker phone Karthik, because everything is echoing.

Vishna, your voice isn't very loud, could you turn the volume up or sit nearer the microphone.

Neervena, I can't see you very clearly – can you see me?

I think we might be able to improve the sound quality if we turn the video off.

Can one of you just check that everyone can hear me and can see the screen? Thanks.

I see that Amit and Anja are now connected – can you two hear me OK?

Some of your voices sound very faint. Are you actually talking into the mike / microphone?

Presenter: checking all participants are present

How many others are we expecting to join?

We'll wait just a few more minutes for the others to join.

OK. I think we are all hear now, so let's start.

Listener: technical questions and comments

Have you shared the screen? I can't see it at the moment.

OK I can see the screen now.

We're just waiting for two more people to join.

Please can you talk up a bit, we can't hear you very well.

This is Milos. I can hear you fine.

This is Olga. I can't hear what you're saying – there's a high-pitched noise.

Establishing ground rules

OK everyone is here now. First could I ask you all to introduce yourselves? Just your name and department will be enough.

We have a couple of people on the call who are not native speakers. If this call is to be successful, we need the native speakers to speak as clearly as possible.

If anyone isn't sure about something please feel free to ask for the information to be repeated or clarified.

Also, can I just remind you all to say your name when you speak. At least the first few times.

And if you ask a question, try and direct it to someone in particular.

Referring to your English

During the demo please feel free to ask questions when you want. But my English is not great, so I would really appreciate it if you could speak very slowly and clearly.

I have a favor to ask you. I am actually used to listening to British people so I am not very familiar with American accents. So if you could speak reasonably slowly that would be great.

If I'm speaking too quickly just let me know. I know have a strong accent, which some people find quite difficult to understand. At any point, bring up any questions you have.

Speaking on behalf of the non native speakers, I would really appreciate it if you could all speak very slowly and clearly.

Checking everyone has the right documentation

Did you all get the files I sent you last night?

Do you all have a copy of the agenda?

Have you all got the presentation open at slide 1?

Do you all have the document in front of you?

19.3 Introductions and agendas

Social chit chat while participants are joining the call

So Praveen, what's the weather like in Bangalore?

Olga, how did the conference go?

Here it is pouring with rain, what's it like with you?

Milos, what time is it with you?

Karthik, how was your holiday?

Yohannes, how are things going in Ethiopia?

Introducing yourself

First of all thank you very much for coming here today. My name's Anh Nguyet and I am a senior ...

Hi, thanks very much for coming. I am Martina. I have been in with ABC for five years. I am a software analyst in the xxx team. It's great to be here and to have an opportunity to tell you about …

I was part of the team that created the new engine, which we are going to look at together.

Most of you already know me through conference calls and emails.

I've been with the company for about three years.

My main role is customer support.

At the moment I am focusing on the core tech development, in particular the Demon project, which is basically the subject of today's session.

I started work as … then I went to work with … So here I am.

Finding out what they already know before you start

Before I begin the main part of the presentation, I'd be interested to hear what you already know about xxx.

OK thanks. So first can you each tell me a little bit about yourselves and how you are currently using x.

Before I start, I would just like to know what features of the product you generally use.

Checking the time they have available

Before I start, could you just confirm how much time you have available.

Giving agenda

So I am going to show you how x works and what you can do with it.

So you'll see how it works and what you can do with it.

The first two parts will take about two hours.

I am planning to have regular coffee breaks.

By the end of the first day you should already be able to do x and y, by the end of the second day you will have become really familiar with it.

I have scheduled a Q and A session at the end, but please feel free to ask me questions at any time.

The second day will be more hands on.

Giving your agenda to small audience (not open to negotiation)

This demo should be interesting / useful for you because …

It's just a short presentation – around 20 minutes.

Before we actually start I just wanted to run through what I had planned for today. So I'm going to start with an overview of X. Then I'll move on to Y, and then … The whole thing should take about 30 minutes. If you want to take notes, that's great, but I will be giving you a handout after the presentation. Also it would be good if you could ask questions at the end. How does that sound?

This is not a demo to teach you how to sell our products to clients. My aim is for you to understand how to use the features.

I am planning to do a general overview of each feature. Then we can do an exercise so that you can see for yourselves how the features work and then you can ask any questions. Some features are identical to the new version, some have been extended, and some are totally new. So when I explain a feature it would be useful if you could tell me what you already know about these features.

If it's OK with you, could you keep your questions until the end, unless there's something that you don't understand.

Giving your agenda to small audience (open to negotiation)

This is the agenda.

Is there anything you think that I don't really need to cover?

Is there anything you would like me to add?

Before we actually start I just wanted to run through what I had planned for today. So I was thinking of starting with an overview of X, because I think most of you are not that familiar with it. Then I'll move on to Y, and then I thought perhaps we could have a short training session. After that we can have a break for lunch, and then continue the training … The whole thing should take about four hours. Does that sound OK, or are the some things you want to leave out or do in a different order?

What to say if agenda looks rather long

Don't worry. We won't be …

… going into great detail.

… discussing all of this.

… spending more than a couple of minutes on each item.

Checking if they need explanations of any key terms

If I use any term that you don't understand raise a hand. Are you familiar with the term 'ticker'?

Referring to the handout

I have prepared some notes on some fundamental definitions of x, y and z. While we are waiting for the others to arrive, could you just have a quick look through them and ask me for any clarifications.

This part of the presentation is actually contained in the handout. So I am just going to go through it very quickly. I will be giving you two examples – x and y. In any case, I will tell you which parts of the handout you need to look at in more detail later.

19.4 Checking and clarifying

Presenter: Checking that participants know where they are in your demo

So, I am going to move on to the next slide now, which is slide 12.

So, we are on slide 12 now. I'd like you to focus on the figure at the top left. The one that says 'functionality'.

Can we just go back to the top of page 20.

OK, so is everyone on page 40? The middle of the page where it says 'How to set up version 2'.

Participants: Interrupting

Sorry, I am not sure who is talking. Can I suggest that everyone announces who they are before they speak?

This is Olga again. I'm sorry but it's hard to understand two people talking at once.

Sorry, but the line isn't great at my end, could you all speak more slowly?

Sorry, what slide are we up to?

Sorry, what page are we on now?

Sorry, I am not sure which figure you are talking about.

19.5 Demonstrating / giving tasks

Giving instructions

So what I want you to do is …

So what we are going to do is …

I would like you to do x.

Please can you …

Explaining reasons for doing the task

The reason for doing this exercise is …

The benefits of you doing this task are …

After you have done this task you will be able to …

This is important because …

Demonstrating / Explaining how to do something

So let's look at how you do it.

Let's imagine that this is the layout you have just created. You can then …

First, you need to have x. After this you can create y. And then you can do z.

The first thing to do is …

If I click here, I can …

I'll hit the pull-down here.

I'm back where I started.

Explaining where things are located

On the left hand side here …

Down here in this section …

Over here is …

You right-click on here.

Explaining how to do something

There are two ways to do this:

> to create a new menu bar you need to …

> to change this menu, go to …

Explaining what they can do

If you press the shift key you can …

This feature allows you to …

You can change its position in the queue.

As you can see …

As you know …

You can only manage x via the interface.

You can redefine which users can use this file.

As with a normal table you can …

So let's do an exercise now.

Dealing with technical problems

Sorry about this …

It seems to be taking a long time to upload.

It's a bit slow at loading.

Do any of you know why this is happening?

I have never had this problem before.

Shall we just have a break while I try to sort out the problem.

19.6 Talking about enhancements to products and services

Introducing the topic of enhancements

So I thought we could begin by looking at the enhancements we have made. And then I have some great new ideas to put by you to see how they might be useful for your …

So let's look at the enhancements one by one, then you can tell me if there is anything you would like to add, and I can make suggestions.

I am really happy to have this opportunity to take you through some other enhancements that I am sure you will find particularly useful.

Walking the participants through the enhancements and getting them interested

This one I think you'll find is really interesting …

The next one should solve a lot of problems for …

We've designed this one because it will significantly simplify / speed up the process of …

Comparing old with new version

Whereas in version 3.5 there is an x, in the new version there is a y.

Another new feature is …

In the new version x has been replaced with y.

Some of the tables that you had in 3.5 no longer exist in 5.1.

X is similar to 3.5, but Y is quite different.

In this version there is no x, so to do y you have to do z.

19.7   Asking the audience questions

Checking audience's understanding

Is there anything you'd like me to go through again before we move on? Do you have any questions at this point? Right. Let's move on to the technical aspects.

Everybody with me?

Any questions on that? Because that's a fundamental assumption.

Any questions up to this point?

I'm going through this pretty quickly. Am I going too fast?

Is there anything you would like me to review?

I'll just give you a quick overview of ...

Would you like me to give you a few more details about this?

You're very quiet - give me some feedback.

I know that some of you may not have had that much experience with x. Which points would you like me to go through again?

Asking open questions to learn about audience's / customers' usage of product or service

How do you think this feature would be useful for you?

Alright, so I know you like this feature, but how could it be optimized?

How do you think your customers might benefit from this?

What other features do you think customers might like?

To what extent would this resolve the problems they are having at the moment?

How do you think they might react to these changes?

What would be the best way to achieve this do you think?

I really value your input on this one – so what suggestions do you have?

Could I ask you a favor, do you think you could ask your customers about this one, and then get back to me?

OK, what about this spreadsheet? Do you like it, is it useful, or would it better to replace it with an excel sheet?

How useful would it be if we replaced it with an excel sheet?

What are the benefits of x?

Which part of this do you use the most and why?

What features here have you never used? Why not?

What new features would you like to see here? Why?

Which of these features would you not like to lose? Why?

Was this what you were expecting? Why (not)?

What has your experience been with this? Have you found this feature useful?

What if we did x here?

How would you improve x?

Checking your understanding of the audience's question

Audience: So is it possible for you to remotely access what we are doing?
You: So you want know if we can access you remotely, is that right?

So you are asking if … Have I understood you correctly?

Sorry I didn't quite catch that.

Could you say it again a bit more slowly?

Could you just repeat the last part again?

I am not sure what you mean by 'widget'.

Sorry I'm not absolutely sure about the answer.

I'll liaise with my colleagues and get back to you.

Making comments on what they say, showing interest

So you've been using it for over a year, I see. And what features do you like most?

Oh, so you haven't actually used it much. Can I ask why?

That's great.

I see.

Right.

Does that answer your question?

When you can't answer a question

Can I get back to you on that?

It will take me a while to answer that – would you like to ask me afterwards?

I'm sorry I can't answer that question, I'll have to check with a colleague but I'll make sure I get back to you by Friday / within the next two days.

When you have spent too long answering questions

I am worried that we won't have time to do everything so I'd like to get back to the presentation. So I'm going to skip a few slides, but don't worry because they are all on the handout.

19.8   Managing the day's events, concluding

Suggesting a break

OK, so we've finished looking at …, would now be a good time to have a 10-minute coffee break?

This might be a good time for a break – what do you think?

OK I think we could have a break now. Let's meet back here in 15 min, that's at 10.20.

Do you think you could possibly be responsible for getting people back here? Thank you.

Could you tell me where the toilet/bathroom/restroom is?

Is there anywhere where I can get a bottle of water?

No, thanks I won't have any coffee myself – I just need to make a phone call.

When you have 10 minutes to go on the first day of a two-day presentation

So we've got about 10 minutes left. I'd just like to summarize what we've done today, so if you've got anything you want me to go over again we can look at it tomorrow morning.

If any of you need to go straight away please feel free to so, but if you have any questions please feel free to stay.

I am sure you will all want to go home now, but could you just ask any questions now so that I can prepare answers for you for tomorrow.

End of first day

Well I think we've covered all the main points. But just before you go, if you have any questions now is the time to ask them. Then I'll either give you a quick answer now or get back to you later.

OK, that's all we've got time for today. So, I'll see you here again tomorrow morning.

OK I'll see you all tomorrow at 9.0.

Have a nice evening.

When you have 10 minutes left on the final day

So that just about winds things up. We've got about 10 minutes left if you have any more questions. OK, well thanks very much for coming. I hope this was useful. If you have any feedback I'd be very grateful to hear it – I think you already have my email address.

Summarizing agreements made or next steps planned

So we've agreed that you will … and that you will get back to me with …

If it's OK with you, I will send you a reminder in 10 days.

So, the next step for you is to implement what you have learned today and …

We didn't have the chance to do all the exercises and tasks, so if you could finish them off by yourselves and then send them to me.

End of final day

If anyone has any further questions I would be happy to answer them after the presentation.

You can find this demo on our website at www.etc. Once again, thanks for coming – I hope it has been useful.

I hope you have found this demo interesting. If you need further details or would like a copy of the presentation then …

Saying goodbye

OK, well I think that's it. It's been great working with you and I hope to see you again some time.

Well thank you very much for coming.

Thanks everyone for making this call, particularly you Karthik, it must be in the middle of the night for you!

Thanks for your time everyone.

Bye everyone.

See you next week.

20 USEFUL PHRASES: FORMAL BUSINESS PRESENTATIONS

20.1 Introductions

Introduction via conference call / webcast

Good afternoon to those of you listening in *place*, and good morning to those in *place*.

It gives me great pleasure to welcome you to …

I am delighted to welcome so many *type of person* here today and we appreciate your interest in *topic of presentation*.

Introducing speaker / s

Joining me today are *names of people*.

I would just like to begin by introducing the team this morning.

To my right, your left, *name and position*. Next to him / her is *name and position*. Then to my left, your right, is …

And in case you don't know me, I'm …

I'd now like to introduce *name*.

Introducing yourself: example 1

Just to give you a bit of my own background.

I did my Ph.D. / MSc / degree in *subject* at the University of *name*. I then did a post-doctorate in *subject*.

I started my career at *name of Company 1*, working on …

I then moved on to *name of Company 2*, where I …

I joined *name of current company* in 2013 as a *position*. I then became *current position* in 2015.

The rest of the team here today also has a lot of experience, not only in *field*, but also in *field*.

Introducing yourself: example 2

Good afternoon. My name is *name* and I'm the *current position* at *current company*.

A. Wallwork, *Presentations, Demos, and Training Sessions,*
Guides to Professional English, DOI 10.1007/978-1-4939-0644-4_20,
© Springer Science+Business Media New York 2014

234

I've been in the role for three years, responsible both for x and y.

Prior to this role I was based in *name of country* for three years.

You've heard from *name of colleague* about … So now I'm delighted to have the opportunity to talk to you about … and how we …

20.2 Agenda

Agenda (very formal)

In a moment I shall say a few words on … *Name of person*, our technical director, will then update you on … and talk about … After that, we will move to the formal business, and I will invite you to put forward any questions you may have.

Today's presentation covers five topics. These are: First …

After our prepared remarks, we will be pleased to take your questions.

Agenda (neutral)

Today I'll cover:

why we believe there are interesting opportunities for …

why we think we are the most attractive partner in terms of …

what outcomes we have set ourselves for …

First let's look at …

20.3 Describing your company

Field of business and structure

As you know, we are in the field of …

We have been working in that field since … though originally we were in the field of …

We are a subsidiary of …

We are the parent company of a group comprising …

We are affiliated with …

We have 100 branches throughout Europe.

We operate in Europe, the US and Asia.

Our company employs 1500 people worldwide.

History and future prospects

Thank you for the introduction and the opportunity to tell you something about *company name.*

The company was incorporated in 2006. We are based in Palo Alto. In 2014 we decided to go public.

Our current turnover is $ 000 and for the last three years we have had net profits of around $ 000.

We have two different divisions. One division focuses on … Our second division is …

All of our technology is patent protected and is based on our knowledge of how to …

We have an experienced management team, and we also believe that we have a very special business structure, which has allowed us to …

A quick look at the market breakdown shows that …

We are excited about the growth prospects for …

Now I would like to give you a quick tour of …

Financial position

This has been a very exciting year for us at *company.*

Not only has … but also …

The business achieved a further year of operating profit growth, up 5 % to $ 34 million. Earnings increased by 2 %, to £ 13 million …

This was despite … This resulted in … Overall we were able to …

Last year was a challenging year, next year will be no less so,

I am pleased to report that …

We have also continued to strengthen our …

Looking ahead …

Our priority is now to …

20.4 Transitions and conclusions

Moving on the next stage in the presentation

I'd now like to pass the presentation to *name* for some closing remarks.

I will now hand over to *name of person*, who will update you on …

That concludes our prepared remarks. We'd now be pleased to take your questions.

Concluding

You have heard a lot today but it can be summed up in these three points:

Thank you very much. I would like to hand you back to *name of co-presenter*.

Thank you for your questions. I am now going to bring the formal part of our conference to a close.

THE AUTHOR

Adrian Wallwork

I am the author of over 30 books aimed at helping non-native English speakers to communicate more effectively in English. I have published 13 books with Springer Science and Business Media (the publisher of this book), three Business English coursebooks with Oxford University Press, and also other books for Cambridge University Press, Scholastic, and the BBC.

I teach Business English at several IT companies in Pisa (Italy). I also teach PhD students from around the world how to write and present their work in English. My company, English for Academics, also offers an editing service.

Editing and proofreading service

Contact me at: adrian.wallwork@gmail.com

Link up with me at:

www.linkedin.com/pub/dir/Adrian/Wallwork

Learn more about my services at:

e4ac.com

A. Wallwork, *Presentations, Demos, and Training Sessions,*   237
Guides to Professional English, DOI 10.1007/978-1-4939-0644-4,
© Springer Science+Business Media New York 2014

Index

This index is by section number, not by page number. Numbers in bold refer to whole chapters. Numbers not in bold refer to sections within a chapter.

A. Wallwork, *Presentations, Demos, and Training Sessions,*
Guides to Professional English, DOI 10.1007/978-1-4939-0644-4,
© Springer Science+Business Media New York 2014